Environmental problems and their management present one of the main areas of concern for the Soviet government in the 1990s. In this original and perceptive study, Professor Philip Pryde examines the pervasive nature of biosphere disruption and environmental contaminants in the country and the extent to which they are damaging the Soviet populace and the resource base upon which it depends. The author also analyzes changes in both public and government perceptions of and responses to environmental problems in the Soviet Union and he assesses possible future trends in Soviet environmental management. Topics given specific attention include air and water pollution, energy resources, toxic waste, biotic conservation, national parks and landscape preservation.

This book is the first comprehensive study of Soviet environmental issues and is written by an international specialist on Soviet environmental problems and management. It is also a most timely publication. In recent years the Western world has been following developments in the Soviet Union with great interest and environmental protection has become one of the primary topics of *glasnost'*. *Environmental management in the Soviet Union* will therefore become essential reading for Western and Soviet environmentalists, and for students and specialists of conservation and environmental management with a special interest in the Soviet Union.

ENVIRONMENTAL MANAGEMENT IN THE SOVIET UNION

Cambridge Soviet Paperbacks: 4

Cambridge Soviet Paperbacks is a completely new initiative in publishing on the Soviet Union. The series will focus on the economics, international relations, politics, sociology and history of the Soviet and Revolutionary periods.

The idea behind the series is the identification of gaps for upper-level surveys or studies falling between the traditional university press monograph and most student textbooks. The main readership will be students and specialists, but some "overview" studies in the series will have broader appeal.

Publication will in every case be simultaneously in hardcover and paperback.

Books also published in this series

Soviet Relations with Latin America, 1959–1987
NICOLA MILLER

The Soviet presence and purposes in Latin America are a matter of great controversy, yet no previous serious study has combined a regional perspective (concentrating on the nature and regional impact of Soviet activity on the ground), with diplomatic analysis, examining the strategic and ideological factors that influence Soviet foreign policy. Nicola Miller's lucid and accessible survey of Soviet–Latin American relations over the past quarter-century demonstrates clearly that existing, heavily "geo-political" accounts distort the real nature of Soviet activity in the area, closely constrained by local political, social and geographical factors.

Soviet policies in the Middle East: From World War II to Gorbachev
GALIA GOLAN

This is the first comprehensive study of Soviet interests in the Middle East. Concentrating on policy developments, Professor Golan analyses major Soviet decisions and objectives. She presents a series of broadly chronological case studies including an examination of the main Soviet alliances: Syria and South Yemen; Sadat's Egypt and Khomeni's Iran. Specific attention is given to Soviet attitudes to the Arab–Israeli conflict, and the role of communism and the importance of Islam in Soviet–Middle East relations. The author concludes by analysing Gorbachev's interests, initiatives and "new thinking" on the Middle East.

Gorbachev in power
STEPHEN WHITE

President Gorbachev's administration has captured the imagination of the world and this book provides the first comprehensive and up-to-date account of the initial five years of Gorbachev's administration. Stephen White examines Gorbachev's political reforms and the process of democratisation; his commitment to *glasnost'* and the extent to which Gorbachev's economic reforms have been put into practice. He also explores nationality questions and the changing role of the Soviet Union in international affairs. A final chapter places Gorbachev's administration within the wider context of the politics of *perestroika*.

Environmental management in the Soviet Union

PHILIP R. PRYDE

Department of Geography, San Diego State University

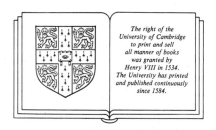

The right of the
University of Cambridge
to print and sell
all manner of books
was granted by
Henry VIII in 1534.
The University has printed
and published continuously
since 1584.

CAMBRIDGE UNIVERSITY PRESS

Cambridge
New York   Port Chester
Melbourne   Sydney

Published by the Press Syndicate of the University of Cambridge
The Pitt Building, Trumpington Street, Cambridge CB2 1RP
40 West 20th Street, New York, NY 10011, USA
10 Stamford Road, Oakleigh, Melbourne 3166, Australia

First published 1991

Printed in Great Britain at the University Press, Cambridge

British Library cataloguing in publication data

Pryde, Philip R. (Philip Rust)
  Environmental management in the Soviet Union – (Cambridge
  Soviet paperbacks).
  1. Soviet Union. Natural resources. Management
  I. Title
  333.7150947

Library of Congress cataloguing in publication data

Pryde, Philip R.
  Environmental management in the Soviet Union / Philip R. Pryde.
    p.   cm. – (Cambridge Soviet paperbacks: 4)
  Includes bibliographical references.
  ISBN 0 521 36079 X. – ISBN 0 521 40905 5 (pbk.)
  1. Environmental policy – Soviet Union. 2. Environmental
protection – Soviet Union. I. Title. II. Series.
HC340.E5P79 1991
363.7′056′0947 – dc20 90–1855 CIP

ISBN 0 521 36079 X hardback
ISBN 0 521 40905 5 paperback

This book is dedicated to all those in the Soviet Union who have toiled, often against great odds, to instill an environmental consciousness in the conduct of that nation's affairs, and to create a more stable global environment for the benefit of all its citizenry and their descendants.

Contents

Illustrations

Tables

Foreword

Zeev Wolfson

Environmental management in the Soviet Union is appearing at a very opportune moment. It is timely not only because in recent years the western world has been following developments in the Soviet Union with great interest, but also because environmental protection has become one of the primary topics of *glasnost'*. Moreover, a book about ecological problems in the Soviet Union would be timely at any point in the last twenty to thirty years for the simple reason that environmental disasters developing on a territory equal to one-sixth of the entire world land mass cannot help but influence the ecological balance of the entire planet. (For a more accurate calculation, it would be necessary to add to this sixth the territory of those other countries whose ecological policy was formed under the influence of the same ideological and economic principles followed in the USSR).

Environmental management in the Soviet Union reaches the reader at a timely moment also because there is now greater hope and greater opportunity for effective cooperation between West and East to mitigate the global ecological crisis. In light of these opportunities, the publication of this volume will be useful to both western environmentalists and their Soviet colleagues. The work gives the reader a broader and more objective analysis of the ecological situation in the Soviet Union than is often found in other western or Soviet publications.

In recent years, Soviet specialists have been publishing more and more analytical works that deal with the most topical aspects of environmental management. They now use much more complete and accurate data about environmental pollution and comment more freely about the effectiveness or ineffectiveness of methods of combating it.

It is no accident, moreover, that recently the Vice-President of the Academy of Sciences of the USSR, V. Koptiug, called on ecologists to switch from a consideration of local situations to a systematic analysis of problems in the country as a whole. Many Soviet works still lack breadth and in several of them emotions of *glasnost'* and *perestroika* dominate over systematic scientific analysis.

The author of this volume, Philip R. Pryde, and the two contributing authors, Dr. Kathleen Braden and Dr. Philip Micklin, combine precisely these important qualities – breadth in the analysis of the problems on the one hand and scientific objectivity on the other. In the space of a little under 300 pages they create a vast picture of environmental management in the Soviet Union from energy production, including atomic energy, to nature reserves, to environmental planning and the nature of public participation in it. The book focuses on the near future, when we can probably expect growing conflicts of both domestic and international scope over sources of fresh water, the transfer of pollutants from source regions to others, and the shortage of many natural resources. Summing up the current situation, the book concludes with a chapter looking ahead to year 2000.

Environmental management closely merges with ecological policy, and the latter is becoming a political issue in all senses of the word. Consequently, one can expect ecological problems to affect other policy fields including the military. Can nations prevent environmental problems from turning into political confrontations in the international arena? If this is possible, then an understanding of the commonality of problems shared by East and West will play a primary role in bringing it about.

This volume should enable the reader to understand not only *what* is happening in connection with various environmental issues in the Soviet Union, but also the background of these processes. It should help the reader comprehend the complexities and opportunities for cooperation between East and West in the ecological sphere. Hopefully, this work will clarify many topics to both ecologists and people in other specialties who by profession, conviction or in a burst of despair or hope are involved in common efforts for the survival of mankind.

Zeev Wolfson, under the pseudonym Boris Komarov, authored the influential work *Destruction of Nature in the Soviet Union*.

Preface

The rapid pace of events in the Soviet Union in the late 1980s, symbolized by the concepts of *glasnost'* (candor) and *perestroika* (restructuring), produced fundamental changes in virtually all aspects of Soviet society. Greater change took place in the four years from 1986 through 1989 than had taken place in the forty that preceded. Certainly environmental issues were no exception to this. A basic purpose of this work will be to chronicle and analyze changes in both public and governmental perceptions of, and responses to, environmental problems in the Soviet Union, and to give thought to probable future trends and needs in Soviet environmental management.

With regard to the nature of changes in Soviet environmental practice that have taken place in recent years, it was natural that the main benchmark for comparison was the author's earlier work *Conservation in the Soviet Union*, published in 1972 by Cambridge University Press. One of the most immediately apparent differences in the two works is the emergence of entirely new topics, some requiring their own separate chapters, in the current volume. Among these "new" environmental topics are nuclear energy, toxic wastes, and environmental cooperation, and of course the environmental ramifications of *glasnost'* and *perestroika*. Not that any of the "old" environmental problems have gone away; indeed, most of them are at least as acute today as they were two decades ago. Thus, the inescapable conclusion arises that Soviet environmental problems have become more profound, a conclusion shared equally within the USSR and abroad, and lends additional timelessness to this study.

The present work will also embody a slightly different emphasis than the earlier *Conservation in the Soviet Union*. The primary focus of that work was upon the Soviet natural resource base, how efficiently it was being utilized, and on the adverse ways in which waste, mismanagement, and pollution affected it. Although in places similar themes appear in the present work, the main emphasis here will be on the pervasive nature of biosphere disruption and environmental contaminants, the extent to which they are damaging the Soviet populace and the resource base upon which it depends, and the effectiveness of the Soviet response to date in dealing with these problems.

It is hoped that this book will be of value to a wide audience of readers, and to that end a selected, but nonetheless fairly extensive, bibliography is presented. An effort has been made in the bibliography to include among the entries a significant percentage of relevant works in English, to assist readers not familiar with the Russian language who desire additional information. The bibliography is necessarily selective, particularly with regard to the wealth of secondary source materials that has emerged in the current *glasnost'* era. At least as many works again as appear in the bibliography were reviewed in some way in the preparation of the various chapters.

The transliteration system used is in general that recommended by the US Board on Geographic Names. For simplicity in the text, however, the soft sign (usually represented in English by an apostrophe) has been omitted from place names (e.g., Gorkiy rather than Gor'kiy), but retained in personal names (e.g., Vorob'yev) and in transliterated Russian words (e.g., *glasnost'*).

My thanks go out to the many persons who assisted in the preparation of this volume. A particular debt is owed, and duly extended, to the two specialists who were kind enough to prepare chapters 7 and 12. Dr. Kathleen Braden is a widely published researcher on Soviet forestry practices, particularly as they relate to environmental issues, and is co-author of *The Disappearing Russian Forest*. She is also active in joint US–USSR efforts to preserve the snow leopard. Dr. Philip Micklin has likewise published extensively on Soviet water management problems, and is the leading western authority on the subject of Soviet river diversion proposals. His research has also included many trips to the USSR to work firsthand with Soviet specialists. Their generous donation of time and talent on behalf of the book is much appreciated. A similar expression of thanks for writing the book's foreword goes to Zeev Wolfson, who, under the pen name of Boris Komarov earlier authored the widely cited *Destruction of Nature in the Soviet Union*. His current journal *Environmental Policy Review* has been an especially helpful resource on contemporary Soviet environmental problems.

Appreciation is also extended to the several specialists who assisted by reviewing drafts of various chapters. These included Brenton Barr, Andrew Bond, Kathleen Braden, William Dando, Richard Little,

Nancy Lubin, and Zeev Wolfson. Their comments and suggestions were most helpful. Additionally, special thanks go to Kathleen Braden, R. A. French, Andrew Bond, Zbigniew Karpovicz, and William Freeman for granting the author the privilege of using their personal libraries or research materials. Also very helpful were personnel at the various libraries where research for this volume was conducted, including the Library of Congress, libraries at the University of California campuses at Berkeley, Davis, and San Diego, as well as at the University of Washington, San Diego State University, the Universities of Birmingham and Glasgow in the United Kingdom, and at the IUCN's World Conservation Monitoring Center in Cambridge.

A similar expression of thanks is extended to the many colleagues, both in the West and in the Soviet Union, whose spoken insights on Soviet environmental problems lent greater accuracy to this work. Others who assisted the preparation of the book in various ways include Barbara Aguado and Sharleen Lane, whose cartographic talents have been especially appreciated, as well as Debra Turner, Jacqueline Nicol, and Teri Fenner who assisted with the manuscript. The author, of course, accepts full responsibility for any errors of either a factual or interpretive nature that may not have been excised from the final draft.

Some portions of certain chapters contain material which, in an earlier form, appeared previously in various journals. Permission to reprint this material is gratefully acknowledged. Specifically, appreciation is extended to V. H. Winston and Sons, Inc., publishers of the journal *Soviet Geography*, for permission to use in revised form material that appeared in issues 1984, no. 1, 1984, no. 6, and 1988, no. 6; to the Marjorie Mayrock Center for Soviet and East European Research, publishers of *Environmental Policy Review*, for permission to reprint material from the June 1987 issue that appears in this volume as Table 14.3; to Elsevier Publishing Co., for permission to use in revised form materials that appeared in *Biological Conservation*, 1986, pp. 351–74, and 1987, pp. 19–37; to the *Current Digest of the Soviet Press* for permission to quote excerpts from their publication; and to Lynne Rienner Publishers, Inc., for permission to use revised materials that appeared in *Environmental Problems in the Soviet Union and Eastern Europe*, edited by Fred Singleton. Last but not least, appreciation is extended to the United Kingdom Atomic Energy Authority for its permission to reproduce Figures 3.3 and 3.4. The author's thanks are also extended to the Board of Regents of the University of Wisconsin System for permission to use revised versions of materials previously appearing in *Soviet Geography Studies in Our Time: A Festschrift for Paul E. Lydolph*, edited by Lutz Holzner and Jeane M. Knapp, the University of Wisconsin-Milwaukee, 1987.

1 Introduction: The state of the Soviet environment in the 1990s

If the October Revolution of 1917 is the most important event in the history of the Soviet Union, the second most important single date may prove to be April 26, 1986. For on that date, not only did a nuclear reactor blow up, not only did one of the most widespread environmental disasters in history take place, but that date also marks the probable point at which a majority of decision-makers in the Soviet Union became convinced that far-reaching, fundamental changes in the way the country had been governed for the previous sixty years had to be enacted. It may have been the day when the already adopted-in-principle transformation that General Secretary Gorbachev terms the "revolution" of restructuring (*perestroika*) became an urgent and generally accepted concept, rather than just an intellectual one. An environmental disaster, with its vast economic repercussions, had forced the issue; the Soviet body politic realized it had no choice but to radically reform the way it governed the country.

To be sure, long before Chernobyl and the much-publicized *glasnost'* (openness, candor) policy of the Gorbachev era, the Soviet Union was willing to admit to a certain amount of environmental pollution. These shortcomings were usually attributed to middle-level mismanagement within the bureaucracies of the various ministries, and it was always assumed that they were curable with increased administrative vigilance and dedication. This assumption became highly questionable even within the USSR by the revelations of ministerial inefficiency and corruption that emerged following the onset of the *glasnost'* era in 1986.

Since the early 1970s, numerous books have appeared in the west on the subject of the Soviet approach to environmental problems (Pryde, 1972; Goldman, 1972; Fox, 1971; Jackson, 1978; Volgyes, 1974; Singleton, 1976; Shabad and Mote, 1977; DeBardeleben, 1985; Ziegler, 1987; Jancar, 1987; Weiner, 1988; and others). In 1978, an outspoken and insightful critique of Soviet environmental practices was published by a well-placed Soviet official, now an *émigrée* (Komarov). In addition, numerous Soviet works have also appeared on the subject of environmental management (e.g., Blagosklonov, Inozemtsev, and Tikhomirov, 1967; Vorontsov and Kharitonova, 1977; Milanova and Ryabchikov, 1979; Nikitin and Novikov, 1986, and many others). To help disseminate the Soviet point of view on environmental issues, the USSR has translated several general works into English (e.g., Frolov, 1983; Gerasimov, 1983; Kolbasov, 1983, 1987; Astanin and Blagosklonov, 1983; Maksakovsky, 1983; and others). As a result, Soviet views and efforts as regard environmental problems are much better known today than they were twenty years ago, as are the difficulties and magnitude of the problems themselves.

The history of Soviet conservation efforts is a most fascinating study in conflicting philosophies and implementation strategies, and has been well reviewed in previous publications. Rather than be repetitive here, the interested reader is referred to Pryde (1972) for a general summary, and to appropriate sections in other works such as those by DeBardeleben, Jancar, and Ziegler. For an excellent detailed examination of conservation efforts in the early days of the Soviet state, and its unhappy history during the Stalin era, the recent volume by Weiner (1988) is an invaluable source.

The state of Soviet environmental quality in 1990

One of the dominant themes of the current *glasnost'* era has been that the state of the Soviet environment is in far poorer shape than has been previously admitted. The Chernobyl nuclear accident forced the more candid public assessment, but so many other problem areas were known to exist that an admission of grave environmental difficulties was inevitable. In 1989, the State Committee for Environmental Protection (acronym: Goskompriroda) published its first "State of the Soviet Environment" report, and in it stated that there were across the country "290 areas of unfavorable ecological situations, totalling about 3.7 million sq. km, or about 16 percent of the USSR" (Doklad, 1989, p. 135). Indeed, the director of the Institute of Geography of the USSR Academy of Sciences has termed one of the worst situations, the Aral Sea problem (see chapter 12), as " a region of ecological calamity" (Kotlyakov, 1988). Similar unhappy assessments were made in the late 1980s by both Yuri Izrael and Fyodor Morgun, both of whom are past heads of environmental protection in the Soviet Union (*Pravda*, Sept. 7, 1987, p. 4; Sun, 1988), and by others (Yablokov, 1990).

Significantly, General Secretary Gorbachev is the first Soviet leader to include environmental issues on his short-list of national priorities, as openly voiced in governmental and international forums. There is no compelling reason to doubt his sincerity in these statements. They appear to reflect an honest response to the acknowledged existence of severe environmental deterioration in the USSR (Gorbachev, 1987).

During the period 1970 to 1990, and particularly since 1986, the Soviet Union has committed itself both philosophically and budgetarily to a major improvement in environmental quality, but, in a word, the results have been mixed. Although a large volume of pollution control facilities went into operation in these years, at the same time great difficulty was experienced in keeping up with the pace of new sources of pollution (Kukushkin, 1986; Komarov, 1980; Pryde, 1983b). Annual expenditures on environmental improvement increased, though at a fairly modest (and therefore inadequate) pace during this period, reportedly reaching about 10 billion rubles in the late 1980s (Poletayev, 1987; Morgun, 1989).

During the Gorbachev era, no significant increase in key environmental expenditures was evidenced until 1988. From 1985 through 1987, governmental outlays on behalf of the environment were not significantly ahead of where they were during the Brezhnev era. For example, total expenditures on environmental improvement in 1987 were only about 10 per cent greater than the 1981–85 average, and the volume of new industrial gas scrubbers installed in 1986 and 1987 was actually slightly less than the 1981–85 average (Table 1.1). These figures would not seem to reflect great official concern over an environmental crisis. However, a sharp increase in environmental outlays was reported as occurring in 1988 (Doklad, pp. 167–8).

Air pollution remains a serious problem in many areas. An emphasis on control devices for industry, and the shutting down of some of the worst plants, has helped a few industrial cities, but new industries have added additional burdens to many air basins. Certain selected cities (most notably republic capitals) have reasonably clean air, but in many industrial centers, the concentration of pollutants in the air "periodically exceeds the maximum permissible norms by a factor of 10 or more" (Morgun, 1989). Contemporary Soviet air quality problems, plus the concerns over world-wide atmospheric deterioration, will be looked at in chapter 2.

Much the same is true of water quality. Chapter 5 will review Soviet efforts to improve municipal treatment facilities, and to remove or treat on-site many sources of industrial pollution. Still, even after

Table 1.1   *Soviet environmental enhancement indices, 1976–1987*

	Annual average				
	1976–80	1981–85	1986	1987	1988
Total expenditures on environmental protection (million rubles)	2,165	2,224	2,615	2,416	3,122
Expenditure on "preserving the atmosphere" (million rubles)	190	180	263	244	317
New sewage treatment capacity (million cu. meters/day)	7.4	5.3	7.0	4.6	
Recycled ("closed") water circulation systems (million cu. meters/day)	24.3	24.4	24.6	25.1	
New capacity, industrial gas scrubbers (million cu. meters/hour)	34.5	40.0	38.8	40.3	

Sources: SSSR v tsifrakh v 1987 g., 1988, p. 275; 1988 data from Doklad, 1989, pp. 167–68

twenty years of effort, much work remain to be done. The quality of some major rivers, such as the Ob', Irtysh, Kama, and Don (see Figure 1.1) has reportedly deteriorated (Poletayev, 1987), and some of the major lakes of the USSR (Ladoga, Aral Sea) have turned into major crisis areas during the past two decades.

Chernobyl has plunged Soviet policy into a state of great confusion. In order to reduce emissions from burning fossil fuels, some in the USSR advocate developing nuclear energy at a rapid pace, Chernobyl notwithstanding. But new coal burning power plants continue to come on line as well, and sulfur emissions continue to create acid rain in the atmosphere. The relationships and trade-offs between energy demand, fuel resources, and atmospheric emissions is a most difficult dilemma that will be examined closely in chapters 2 through 4.

During the 1970–90 period, the USSR has almost doubled its nature reserve network, and has created new systems of national parks and biosphere reserves. It has also enacted a national wildlife law and produced its first Red Books of endangered species (Borodin, 1985). But many threatened species are not adequately protected by the existing preservation network, and both poaching and "economic impera-tives" seem to remain well-entrenched obstacles to effective biotic management (chapters 8 through 10).

Only modest progress had been made in using natural resources

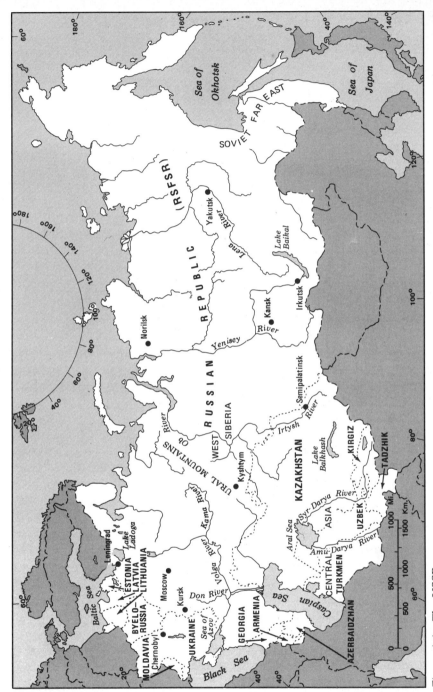

Figure 1.1 The USSR

more effectively. Logged areas are required to be reforested but compliance is spotty at best (chapter 7). The waste that used to be widespread in the petroleum and hard-minerals industries has been reduced somewhat, but is far from having been eliminated. Recycling programs have been started in many areas, but their effectiveness has been marginal (chapter 6). Letters complaining of residual problems in all these areas are still frequently encountered in the Soviet press.

Conservation education programs have been expanded, and the Soviet Union has lately become much more active in the international arena in working for environmental improvement. A new phenomenon in the USSR is the emergence of active public conservation groups. Chapters 14 and 15 will examine these trends, but will also suggest that, with the exception of the Baltic republics and a few other local centers of outspoken environmentalism, conservation still seems to be perceived by the bureaucracy as a *governmental* function, with a reluctance on the part of the ministries to acknowledge that the Soviet people have a decisive role to play, as well.

In short, pro-conservation governmental policies during the 1970s and 1980s have produced many local successes, but the march of new technologies, budgetary restraints, ministerial inertia, and other recurring problems have blunted or negated many of these gains. Achieving a demonstrable level of *net* environmental improvement for the USSR as a whole will be one of the key challenges facing the Soviet Union in the 1990s.

The nature of environmental legislation in the Soviet Union

Environmental management in the Soviet Union has a broad legal framework. Its highest expression is to be found in the Soviet Union's Constitution itself. In the fourth Constitution of the USSR, adopted in 1977, it is specified in Article 18 that "necessary steps will be taken to protect and make scientific, rational use of the land and its mineral and water resources, and the plant and animal kingdoms, to preserve the purity of air and water, to ensure reproduction of natural wealth, and to improve the human environment." Further, Article 42 stipulates that the citizenry has a right to a healthy human environment, and Article 67 states that "Citizens of the USSR are obliged to protect nature and conserve its riches." These constitutional provisions are highly commendable, but do not automatically produce environmental improvements. The quality of a nation's natural environment will be more highly correlated with the degree of success achieved in carrying out its actual environmental laws, than with such generalized constitutional provisions. Unfortunately, Soviet writers themselves are the first to lament the ineffectiveness of these laws, as the ensuing chapters will relate.

The USSR does not lack environmental legislation. At least four main types of environmental enactments can be identified.

The first are Union republic comprehensive conservation laws. These are generalized guidelines, more in the nature of policy statements that provide the broad directives for resource conservation in each of the USSR's fifteen constituent republics. Between 1957 and 1963 all of these Union republics passed such laws.

Secondly, the USSR has enacted several pieces of national legislation, each providing basic guidelines for managing one particular component of the environment. The first of these were Principles of Land Legislation (1968) and Principles of Water Legislation (1970); the text of both these acts is translated in Pryde, 1972. Subsequently, similar acts have been passed relating to minerals (1975), forestry (1977), air quality (1980), and wildlife (1980). A draft law on environmental protection was being prepared in 1989 (*Pravda*, Jan. 17, 1988, p. 1). All legislation and regulations in the area of environmental management that are enacted by subordinate administrative bodies must conform to the principles in these national laws.

Third, the guidelines for each new five-year-plan contain sections dealing with environmental improvements for that five-year period. Although very general in their wording, they are intended to provide policy direction for all the administrative bodies that will have to implement the goals of the new-five-year plan. The environmental protection objectives for the 12th five-year plan are presented in Appendix 1.1. The environmental provisions for the 1991–95 five-year plan, being drafted as this book goes to press, should provide interesting insights into Gorbachev's approach to environmental issues. Many of the individual Union republics also prepare a five-year environmental improvement plan.

A fourth approach involves a large number of special declarations that have been passed by the Communist Party of the Soviet Union (CPSU), the Council of Ministers, or the Supreme Soviet to improve administrative efforts with regard to a particular type of pollution, or to correct problems in specific geographic regions. An early example of the former was a 1949 resolution on combating air pollution (Zeigler, 1987, p. 49); another would be the December 1972 CPSU resolution "On the Intensification of Nature Conservation and Improved Utilization of Natural Resources." Resolutions that are regional in scope include those passed in the 1970s directed at improving water quality in the Volga basin and Lake Baikal; they involve pollution problems which will be explored in chapter 5.

Environmental guidelines that are quantitative in nature are established by the USSR State Committee for Standards. Between 1976 and 1978 this committee established national standards for water bodies, for atmospheric emissions, and for protecting other components of the environment (Ziegler, 1987, p. 112). In general, these standards are strict, sometimes too strict to be enforceable. They will be discussed more in chapter 6.

Research on environmental improvement is carried out by special research and design institutes, as well as by various institutes of the USSR Academy of Sciences. Their work is supervised by the USSR State Committee for Science and Technology, under the Council of Ministers. A Commission on Nature Conservation has existed within the Academy of Sciences since 1955, and, more recently, a Scientific Council on Problems of the Biosphere has been created. Since 1973, the State Committee for Science and Technology and the Academy of Sciences have operated an Interdepartmental Science and Technology Council to analyze and make recommendations for the improvement of environmental efforts in the USSR (Kolbasov, 1983, p. 176). The work of many of the individual research institutes will be touched upon in those subsequent chapters which deal with their specific topics of concern.

The administrative structure of Soviet environmental management

Ultimate responsibility for the state of the Soviet environment, and for the implementation of the foregoing laws, falls to the USSR Supreme Soviet and the Council of Ministers, with environmental policy emanating from the Central Committee of the Communist Party (CPSU) and the Politburo. To assist with these tasks, the Council of Ministers, under its presidium, has established a body known as the "Commission on Environmental Protection and the Rational Utilization of Natural Resources," among whose tasks are the evaluation of proposed environmental laws and exercising the oversight function for implementing these laws once they are enacted.

The overall planning of the development of the Soviet economy, including its environmental improvement components, is the task of the State Planning Committee (generally referred to as "Gosplan"), which is under the Council of Ministers. Within Gosplan there exists the Department for Nature Conservation; its task is to formulate those portions of both five-year and annual plans that deal with natural resource conservation and environmental enhancement (Astanin and Blagosklonov, 1983, p. 142).

The implementation of these plans is the responsibility of the various ministries and corresponding state committees. For example, the Ministry of Fisheries is responsible both for the harvesting of fish according to Plan directives, and for the conservation of commercial and recreational fish stocks. The State Committee for Forestry regulates both the cutting and replanting of forests, and all other aspects of timber procurement (Kolbasov, 1983, p. 185). Other committees deal primarily with planning matters, such as the State Committee for Science and Technology. There exists a large number of state committees, and often their jurisdictions overlap (Jancar, 1987, pp. 56ff). Given that Soviet bureaucracies jealously guard their own empires, the potential for counter-productive squabbling and lack of cooperative efforts is immense.

In 1978, responsibility for monitoring the state of the Soviet environment was centralized when the country's hydrometeorology service was removed from the Ministry of Health, and was elevated, expanded, and restructured as the State Committee for Hydrometeorology and Environmental Control (with the acronym "Gidromet"). Its new tasks included the monitoring of air and water quality in all areas of the country, especially in critical air basins, inspecting factories and construction sites (and temporarily or permanently shutting them down, if necessary), and assessing the degree of compliance with the corresponding environmental regulations and standards (Kolbasov, 1983, pp. 188–89). Although the Committee's charge was quite broad, its focus was mainly on monitoring water and air pollution, primarily the latter (Ziegler, 1987, p. 118f).

With the new assessment of the Soviet environment provided by the Gorbachev administration, a critical review of the work of Gidromet was conducted, and it was found lacking. According to one knowledgeable observer, "new ideas connected to the protection of the environment do not find any special support among the officials of the Committee" (Wolfson, 1988c). However, the inadequacies do not lie entirely within Gidromet; a key problem was that the enforcement function still lay mainly within the various ministries which had been largely the cause of the problems in the first place. Gidromet was not nearly as powerful as it needed to be.

Figure 1.2   Entrance to "Gidromet" in Moscow

It is not surprising that as large a country as the USSR should have a sizable number of agencies involved in matters of environmental protection. On the other hand, it has been frequently suggested by Soviet specialists that perhaps there are too many agencies, too much fragmentation, and too little coordination in the Soviet Union's environmental management efforts. Starting in the late 1970s (Kolbasov, 1983) and continuing throughout the later 1980s (Poletayev and others), there have been many calls for the creation of a larger, more powerful Soviet environmental protection agency. Under Gorbachev's administrative restructuring, this has now been done.

Late in 1987, it was decided to relegate Gidromet back to a scientific research organization, and the words "and the Environment" were dropped from its name. To carry out its environmental management duties, a new agency, the USSR State Committee for Environmental Protection (acronym: "Goskompriroda") was formed. Its creation was part of a major resolution taken by the CPSU Central Committee and the Council of Ministers entitled "On the Fundamental Restructuring of Environmental Protection in the Country." The headline announcing the new agency read "New state committee given broad powers over ministries in coordinating environmental management, issuing waste disposal permits, halting operations that violate norms, suing polluters" (*Pravda*, Jan. 17, 1988, p. 1). It is a Union-republic ministry, which means that each of the USSR's fifteen constituent republics will have a similar local organization.

Its first head was Fyodor Morgun, who appeared to be serious and enthusiastic about his responsibilities, though most of his time was spent simply trying to put together an administrative infrastructure. Morgun was succeeded in 1989 by Nikolai Vorontsov, the first person ever to head a major Soviet ministry who was not a party member. Whether Goskompriroda has been given adequate "power over ministries" remains to be seen.

The 1988 Resolution described the new agency's tasks as carrying out the comprehensive environmental protection activities of the country; environmental monitoring; long-range environmental planning; promulgating norms and standards; overseeing the design, siting, and construction of environmentally sensitive facilities; issuing waste disposal permits; managing nature preserves, endangered species, and hunting; environmental education; and international cooperation and coordination (*Current Digest of the Soviet Press*, Feb. 17, 1988, p. 7f). This is a huge agenda, and will require a powerful agency to implement it. Unfortunately, at the onset of the 1990s it remained very doubtful whether Goskompriroda has been given the necessary funding, authority, and personnel to successfully carry out all these vitally important charges (Freeman, 1989). For example, it operates as only one out of several dozen entities that comprise the USSR Council of Ministers, where it is subject to being consistently outvoted.

Monitoring the effectiveness of environmental laws is carried out by a number of entities, with Goskompriroda now apparently having the primary responsibility. The People's Control Committee, a governmental agency, is another such official organization, though it is unknown if its role has been redefined with the creation of Goskompriroda. It could (if so directed) serve as a watchdog agency at the local level over Goskompriroda's effectiveness.

One other important agency is the Sanitary Epidemiological Service of the USSR Ministry of Health, which has had responsibility for monitoring industrial pollution, particularly as it relates to public health. During the 1980s, the Sanitary Epidemiological Service apparently had the primary responsibility for water quality, rather than Gidromet, but these duties now presumedly lie with Goskompriroda. These and other official oversight agencies (such as local district

Soviets, etc.) are themselves responsible to higher bodies charged with environmental review, such as the Commission for Environmental Protection and the Rational Use of Natural Resources, mentioned earlier, and of course various oversight bodies within the Communist Party (CPSU) itself. Finally, the various republic-level public conservation societies also are permitted a limited monitoring authority; these will be discussed more fully in chapter 14. All of these various administrative and oversight bodies are discussed in some detail in the works by DeBardeleben, Jancar, and Ziegler, to which the interested reader is referred.

At the Union-republic level, parallel regulations and institutions exist. To examine the somewhat varying practices in each of the fifteen republics is not feasible here, but a short summary of the main environmental protection agencies that exist at this level is available in English for the interested reader (Kolbasov, 1983, pp. 190ff).

Coping with the new environmental challenges

The goal of the present volume will be to analyze the state of the Soviet environment at the onset of the 1990s, and to examine in particular the new challenges that must be addressed in the course of the USSR's preparing itself for the realities of the twenty-first century. A major theme of this volume will be that, although the world's environmental agenda has clearly changed and expanded over the past twenty years, there are numerous areas of environmental concern in which the Soviet government (and bureaucracy) has not yet geared up sufficiently to cope with these new challenges.

The complex world of the 1980s and 1990s has forced many new environmental issues to the forefront. The earlier Soviet environmental agenda, which initially evolved in the 1960s, was based on the time-honored principles of using natural resources in the most efficient manner, with an eye to their conservation and, if possible, reproduction (Pryde, 1972). For the most part, the basis of this approach was not *primarily* enlightened ecological principles, but rather involved running the national economy in a more efficient and productive manner. Recent works have suggested that, at least prior to Gorbachev's new emphasis on the environment, this remained the dominant consideration (Bergson and Levine, 1983).

Today there is little doubt that the Soviet Union has awakened to its environmental problems. In fact, one can speak of three distinct periods of "environmental awakening" in the postwar period. The first we might term the "Baikal Awakening," and involved the period from about 1965 through the early 1970s. Focusing on a few large issues such as Lake Baikal, it involved mainly writers and scientists.

The second period may be termed the "Chernobyl Awakening." Extending from 1986 until the early 1990s, it has been characterized by the general public becoming environmental activists, spurred on by fears of nuclear radiation and other issues they see as actually or potentially affecting them personally.

The third epoch can be called the "Perestroika Awakening," and has not happened yet. But it is the most important. It will involve (hopefully) the environmental awakening of the giant Soviet bureaucracy, especially ministry officials and factory managers. Without this third phase, the Soviet environment will remain endangered. It is overdue, and must begin in 1991 with the advent of the 13th five-year plan.

To meet the challenges of the new century, the Soviet environmental agenda needs to become more ecologically based, more widely accepted throughout the bureaucracy, and more international in scope. It is becoming axiomatic that many of the world's most pressing environmental problems are inherently regional, or even global, in nature, and that close cooperation among nations is essential if we are to overcome them successfully. As the world's largest country, the USSR must be in the forefront of advancing this cooperation. Works by Soviet authors that analyze the earth's environmental future have begun to appear, but Soviet foreign (and internal) policy has thus far lagged behind the requisite pace of international cooperation. This theme will be examined in more detail in subsequent chapters.

During the 1970s and the first half of the 1980s, the Soviet conservation agenda had not progressed much from the original precepts of ensuring wise use of natural resources, and attending to the more obvious of the residual pollutants which the utilization of these resources inevitably produced. To be sure, books were starting to appear which evidenced a broader understanding of global problems (e.g., Nikitin and Novikov, 1986), but there was little evidence that this understanding extended much beyond the towers of academia and the halls of the Academy of Sciences. For the ministerial bureaucrats, it was still pretty much business as usual, which meant that fulfillment of the economic plan, and ensuring their own job security, was paramount.

Then, in 1986, this complacency with the conservative, domestic, status quo type of environmental thinking was abruptly shattered.

The Chernobyl accident demonstrated vividly that Soviet environmental problems were world environmental problems. Traditional Soviet xenophobia was forced to give way to much closer international cooperation. A tragic accident had caused the USSR to become irreversibly integrated into the complex sphere of world environmental problems, whether it wanted to be or not.

Chernobyl's significance lies as much in its timing as in its effects. It occurred during the first year of Gorbachev's tenure, almost simultaneously with the inception of his *glasnost'* and *perestroika* policies. If Gorbachev had any doubts in March of 1986 about how strongly he wished to pursue these two new policies, Chernobyl undoubtedly erased them. Chernobyl was the straw that broke the back of the "old" secretive and highly centralized Soviet Union, the final curtain call of the clearly deficient Stalin–Brezhnev model of national management.

The devastating fallout from Chernobyl was the most obvious of the new litany of Soviet environmental concerns, but the 1980s saw numerous other new issues emerge into the limelight. Although most of them lack the immediate newsworthiness of a radioactive cloud, in the long run several of them, such as atmospheric warming and ozone depletion, could prove to be equally ominous. Many of these "new" issues are new only in the sense that people started taking them seriously in the 1980s; problems such as acid rain, toxic wastes, and genetic preservation have their origins a decade or more in the past. It is encouraging that there is now at least a recognition that they must be dealt with in a prompt and forthright manner.

The ensuing chapters will examine these contemporary environmental issues, as well as the more traditional ones, with a particular focus on the effectiveness of the Soviet Union's response to them.

Appendix 1.1
Environmental goals of the
12th five-year plan

Environmental protection and the rational utilization of natural resources

To increase the effectiveness of environmental-protection measures. To introduce progressive technological processes on a broader basis. To develop combined production facilities that ensure the complete and comprehensive utilization of natural resources and raw and other materials and that exclude or substantially reduce harmful effects on the environment.

To consistently improve the protection of the country's water resources. To complete the carrying out of basic measures to protect the basins of the Baltic, Caspian and Black Seas and the Sea of Azov. To continue the implementation of a complex of measures to protect bodies of water in the Arctic Basin, Central Asia and Kazakhstan, as well as to improve the condition of small rivers and reservoirs. To enhance the effectiveness of the operation of water-treatment structures and installations. To expand the use of treated effluents and water from mines for irrigation and other national-economic needs.

To step up the protection of the air. To this end, to perfect technological processes, equipment and means of transportation, to improve the quality of raw materials and fuel, and to introduce highly efficient installations for the purification of industrial and other emissions.

To ensure the rational utilization of land and its protection against wind and water erosion, mud slides, landslides, rising groundwater, swamping, desiccation and pollution. To step up work to improve the conservation of agricultural land and to create shelterbelts. To recultivate about 660,000 hectares of land. To constantly expand the application of methods, harmless to man, of protecting agricultural crops and forests from pests and diseases.

To improve the protection of the earth's interior and the comprehen-

15

sive utilization of mineral resources. To reduce losses of commercial minerals during their extraction, concentration and processing. To ensure environmental conservation in the USSR's economic zone and on the USSR's continental shelf.

To continue the creation, and improve the amenities, of the green-belts of cities and settlements, and to expand the network of nature reserves, national parks, preserves and other protected nature areas. To step up work on the protection, reproduction and rational use of the plant and animal world.

To enhance the effectiveness of state monitoring of the condition of the environment and sources of pollution, and to improve the technical outfitting of this service with effective automatic instruments and equipment. To expand the forms and methods of the participation of public organizations and the population in this work. To instill in Soviet people a sense of high responsibility for the conservation and multiplication of natural resources and their thrifty utilization. To improve the management of environmental protection in the country.

Source: As translated in Current Digest of the Soviet Press, vol. 37, no. 49 (Jan. 1, 1986), pp. 18–19; permission to reprint is appreciated.

2 Managing the Soviet atmosphere

Air pollution has inevitably been the unwanted by-product of industrialization. No nation, regardless of its economic or political system, has been spared this dilemma. The nature of regional air quality problems will vary depending on the local mix of industry, automobiles, and energy production facilities, but in some form, air pollution will be present. In the Soviet Union, more than 60 million metric tons of harmful emissions were released into the atmosphere in 1988 from industry alone (Doklad, 1989, p. 64f). These included 17.6 million metric tons (mmt) of sulfur dioxide, 14.7 mmt of particulates, 14.9 mmt of carbon monoxide, 8.5 mmt of hydrocarbons, and 4.5 mmt of nitrogen oxides. Industry and power plants have historically been the main sources of airborne contaminants in the USSR, but today motor vehicle emissions are also of considerable importance (Table 2.1). The data in Table 2.1 suggest an improvement during the 1980s in total emissions from both stationary and automotive sources. If these figures are complete and accurate, then Soviet emission control efforts (which include shutting down many older, dirtier factories) may be having a beneficial effect.

Industrial and automotive sources

The overriding goal of the Soviet Union's first several five-year plans was the most rapid possible industrialization of the country. Since the nation's economy in the 1930s was relatively small, this meant that the capital available for construction had to be directed towards new production facilities; very little was allocated for pollution abatement. Consequently, in a number of the cities where these early "dirty" plants were built, such as the steel complex at Magnitogorsk, a serious air pollution problem is acknowledged to exist. Since many of these plants employ now-obsolete technologies, this aspect of the problem may not be fully resolved until such time as the older plants are eventually replaced (Mote, 1978).

Table 2.1   *Discharge of harmful substances into the atmosphere*

Source	1980	1985	1986	1987	1988
Stationary sources	72.8	68.3	66.5	64.3	61.7 million tons
Automotive sources	38.0	36.7	37.1	36.2	35.8 million tons
Total	110.8	105.0	103.6	100.5	97.5 million tons

Source: Okhrana, 1989, p. 7

Table 2.2   *Expenditures on air pollution abatement, 1976–1988*

	Annual average		1986	1987	1988
	1976–80	1981–85	1986	1987	1988
Total spent on "Preserving the atmosphere" (million rubles)	190	180	263	244	317
New capacity, industrial gas scrubbers (in million cu. meters/hour capacity)	34.5	40.0	38.8	40.3	n.a.

Source: The USSR in Figures for 1987, Moscow, Statistika Publishers, 1988, p. 275

Yet the situation is far from satisfactory even at newer plants. In 1989 *Izvestiya* reported that of the thousands of emission control devices on factories in the Donetsk region "a significant number of them are never working," and that even the operable ones are unable to contain toxic emissions (the precise toxic gases involved were not specified). Consequently, such toxic emissions exceed air quality norms by ten to thirty times in Donetsk, Gorlovka, Makeyevka, Mariupol, and other cities in the Donets coal basin (Lisvenko and Trach, 1989).

Although billions of rubles have been spent to control emissions into the atmosphere, the pace of expenditures in this regard remains inadequate. Outlays for air pollution abatement in 1987 were 244 million rubles ($390 million at the official exchange rate), but this was only 28 percent greater than the average amount being spent a decade earlier, and actually represented a decline from the 1986 level (Table 2.2). A sharp increase apparently took place in 1988, however. The total expenditure, even in 1988, does not seem impressive; in the United States over a billion dollars is spent annually on pollution control devices for motor vehicles alone. Table 2.2 also shows that the average annual rate of installing new industrial gas scrubbers in the USSR was lower in 1986–87 than in 1981–85.

In the past it has been common for emissions from smaller plants to be poorly controlled, and for new factories to be opened (or enlarged) before adequate pollution control devices are installed (Poletayev, 1987). In the early 1980s, in some industries as much as 25 to 40 per cent of all installed gas scrubbers and dust collectors were ineffective or inoperable (Nuriyev, 1983; *Soviet Geography*, Jan. 1988, p. 13). As a result of such inattention to abatement techniques, it has been suggested that the USSR may produce up to twice as much pollution per unit of output as do western industrialized nations (Komarov, 1980, p. 30).

As a result, high air pollution conditions characterize numerous Soviet cities; in the Russian Republic (RSFSR) alone thirty-six cities are listed as having unacceptably high air pollution (Pipia, 1989). Some of these cities, and others, that have been identified in recent years as having a significant air pollution problem are indicated in Table 2.3. In 1987, Yuri Izrael, then head of the Soviet Union's environmental monitoring agency (Gidromet), acknowledged that about 100 Soviet cities suffered maximum concentrations of harmful pollutants that exceeded norms by at least ten times (Izrael, 1987); this same statement was later published with the data referred to in the next paragraph. Cities with steel mills, non-ferrous smelters or petroleum refineries, such as Norilsk, Ufa, Novokuznetsk, Perm, and others appear to have some of the more serious problems.

In 1988, for the first time, the USSR published specified pollutant discharge figures (from stationary sources only) for thirty-three selected Soviet cities, based on 1987 data. These figures are reproduced as Appendix 2.1. The cities among those listed with the greatest total discharge of pollutants were Novokuznetsk, Magnitogorsk, Mariupol (Zhdanov), Chelyabinsk, and Baku. Those with the highest *per capita* emissions were the first four cities mentioned above, as well as Bratsk, Ust-Kamenogorsk, and Dzhambul. The list contains all fifteen of the republic capitals (some of which enjoy relatively clear air), but omits such large cities as Gorkiy, Kharkov, Sverdlovsk, Omsk, Kuybyshev and several others, and certain of these are known to have fairly dirty industries. Additional data were made available in 1989 (see note to Table 2.3). The new statistical handbook which became available in 1990 provided additional comprehensive data (*Okhrana . . .*).

In addition to the missing metropolises, the 1987 list has other

Table 2.3   *Soviet cities known to have a significant air pollution problem*

City	Source of pollutants	Reference(s)
Alma-Ata	Photochemical smog, dust	*CDSP*, July 13, 1983; *Doklad*...
Almalyk	Copper smelter	Shchepotkin, 1987; *Doklad*...
Angarsk	Chemical plants	*Narod. khoz.*, 1987, 576; *Doklad*...
Baku	Oil refineries	Appendix 2.1; *SG*, 1989, 542–4
Bratsk	Aluminum, wood chemicals	*SG*, 1987, 179; Morgun, 1989; *Doklad*...
Chelyabinsk	Steel mills	Appendix 2.1; *Doklad*...
Cherepovets	Steel mills, power plants	Shchepotkin, 1987; *Doklad*...
Dneprodzerzhinsk	Steel mills	Nuriyev, 1983; *Doklad*...
Dzhambul	Industrial pollutants	Appendix 2.1; *Doklad*...
Fergana	"pollutants exceed health norms"	Poletayev, 1987; *Doklad*...
Frunze	Industrial, photochemical pollutants	Appendix 2.1; *Doklad*...
Irkutsk	Industrial pollutants, smoke	Appendix 2.1; *Doklad*...
Kemerovo	Steel mills	Shchepotkin, 1987; Poletayev, 1987; *Doklad*...
Krasnoyarsk	Wood, steel processing, aluminum	Appendix 2.1, 1987; *Doklad*...
Magnitogorsk	Steel mills	Appendix 2.1; *Doklad*...
Mariupol (Zhdanov)	Steel mills	Poletayev, 1987; Appendix 2.1; *Doklad*...
Mogilev	Sulfur compounds	Appendix 2.1; *Doklad*...
Monchegorsk	Nickel smelter	*SG*, 1989, 255; Whitney, 1989
Moscow	Vehicles, power plants, industry	Appendix 2.1; *Doklad*...
Nizhniy Tagil	Steel mills	*Time*, Jan. 2, 1989; Morgun, 1989; *Doklad*...
Norilsk/Igarka	Smelter, pulp mills	Nuriyev, 1983; Bond, 1984
Novokuznetsk	Steel mills	Poletayev, 1987; Appendix 2.1; *Doklad*...
Perm	Petrochemicals, refineries	*SG*, 1987, 608; *Doklad*...
Semipalatinsk	Smelter, power plants	DeBardeleben, 1985, 37; *Doklad*...
Sterlitamak	Petrochemical plants	Wolfson, 1988a
Togliatti	Chemical plants	Shalgunov, 1984
Ufa	Petrochemicals, refineries	Shchepotkin, 1987; Morgun, 1989; Appendix 2.1
Ust-Kamenogorsk	Non-ferrous smelters	Appendix 2.1; *Doklad*...

Table 2.3   *(cont)*.

City	Source of pollutants	Reference(s)
Yerevan	Rubber, chemicals, power plants	Wolfson, 1988a; *SG*, 1987, 608; *Doklad*. . .
Zaporozhye	"pollutants exceed health norms"	Poletayev, 1987; Appendix 2.1; *Doklad*. . .

Sources: In the above references, *CDSP* refers to *Current Digest of the Soviet Press*, and *SG* refers to the journal *Soviet Geography*.

Notes: Air quality in the following cities – Angarsk, Bratsk, Gorkiy, Nizhniy Tagil, Ufa, Volgograd, and Zaporozhye – was characterized in the article listed in the bibliography by F. Morgun as "especially polluted."

For information on specific pollutants and additional cities, see Appendix 2.1.

Late in 1989, Goskompriroda released its first "State of the Soviet Environment" report (Doklad. . ., 1989). In it (pp. 20–22), there appears a list of 68 Soviet cities having "the highest levels of air pollution in 1988." In Table 2.3 above, 24 of those cities are presented and are indicated by the source reference "Doklad. . .". The other 44 highly polluted cities identified in the Goskompriroda report are: Abakan, Arkhangelsk, Barnaul, Chardzhoy, Chimkent, Chita, Dnepropetrovsk, Donetsk, Dushanbe, Groznyy, Kaliningrad, Kamensk-Uralskiy, Khabarovsk, Kiev, Kirovakan, Kommunarsk, Komsomolsk, Krivoy Rog, Kurgan, Kuybyshev, Leninogorsk, Lipetsk, Lisichansk, Makeyevka, Novosibirsk, Novotroitsk, Odessa, Omsk, Orenburg, Osh, Prokopyevsk, Rostov, Rustavi, Severodonetsk, Shelekhov, Sverdlovsk, Temirtau, Tyumen, Ulan-Ude, Usolye-Sibirskove, Volzhskiy, Yuzhno-Sakhalinsk, Zestafoni, and Zyryanovsk. In addition, the 1989 statistical handbook also listed very high readings from stationary sources in Volgograd, Pavlodar, Ekibastuz, and a number of smaller cities (*Okhrana*. . ., 1989, pp. 22–24)

shortcomings. The lack of data from mobile sources (motor vehicles, etc.) is a significant deficiency, and makes some cities appear to have better air than is actually the case. For example, in Appendix 2.1 Moscow is shown as having fairly low levels of carbon monoxide, but this gas comes mainly from automotive sources, and as Moscow has far more cars than any other Soviet city, it is reasonable to assume this figure should be considerably higher than that shown. Indeed, the newspaper *Sotsialisticheskaya Industriya* reported in February of 1989 that levels of both nitrogen oxides and carbon monoxide in Moscow had exceeded norms by 30 percent, and hydrocarbons by 100 percent. The omission of mobile sources, as well as the fact that many of the worst industrial polluters are found in relatively small cities, accounts for the last line of Appendix 2.1 appearing to show that the thirty-three large cities have cleaner air than the country as a whole. Some of them do, but others are worse than shown, and many cities with major polluters are not included. The 1989 statistical handbook gave emis-

Figure 2.1   Aluminum smelter near Irkutsk

sions from mobile sources for the USSR as a whole, but not for individual cities.

As noted, some of the dirtiest industrial plants are found in relatively small "company towns," such as the copper smelter at Balkhash in Kazakhstan, the nickel smelters at the Kola Peninsula towns of Monchegorsk and Zapolyarny, or the aluminum smelter at Shelekhov, near Irkutsk (Figure 2.1). One analysis of the 1987 data also raises the question of whether the figures include only emissions originating in the city proper, for under adverse atmospheric conditions, a given city such as Irkutsk could suffer from major pollution sources found in nearby towns such as Shelekhov or Angarsk (Sagers, 1989).

Emission sources in the Soviet Union are far-flung. One would not expect to encounter air pollution near the North Pole, but it is there. Polluted air is not uncommon today in winter at very high latitudes, often in regions far removed from any likely source. The name that has been given to these unexpected pollutant concentrations is "Arctic haze." Typical components include sulfates and aerosols, high levels of which have been recorded in Alaska, Canada, Greenland, Spitsbergen, and other Arctic locations. No positive source identification for these pollutants has been made as yet, but research on high-latitude meteorological patterns have led to one suggestion that industrial plants in the Urals might be a possible origin for these contaminants (Rahn, 1984).

Other high-latitude point sources of air pollutants exist in the Soviet Union which might also contribute to the Arctic haze problem. These include the copper-nickel smelters and wood-processing plants around Norilsk and Igarka (Bond, 1984). The situation at Norilsk may be improving, as abatement equipment to capture and re-use sulfur previously emitted into the atmosphere has recently become operational (*Soviet Geography*, 1987, p. 779). The previously mentioned nickel smelters on the Kola Peninsula are also possible contributors to Arctic haze, and have also been suggested as a likely cause of acid rain destruction of forests in neighboring Finland (Whitney, 1989).

It is quite probable that other nations also contribute to this problem, but the USSR undoubtedly has the most extensive industrial development north of the Arctic Circle. To date, however, it has published little about this problem. Inasmuch as these pollutants could have some relationship to the polar ozone depletion phenomenon (see below, p. 28f), Soviet scientists should assume a leading role in Arctic haze research.

Since Soviet citizens collectively own and drive far fewer automobiles than do their West European and American counterparts, automotive sources of air pollutants are at present generally less of a problem in the USSR than in other advanced countries. As noted earlier, however, the city of Moscow may be an exception. The Soviet Union has taken the important step of phasing out lead as a gasoline additive, and some vehicles have been converted to burn natural gas. But as automobile production is rising rapidly in the USSR, there is still a potential for increased automotive pollution in the future, especially in light of an inadequate road infrastructure. Perhaps of more immediate concern is the Soviet Union's huge fleet of heavy-duty trucks whose emissions are very poorly controlled. Indeed, during one inspection of diesel trucks entering Moscow, *Pravda* reported that half of them were producing too much smoke for it to be accurately measured (*CDSP*, 1977, no. 47, p. 5).

In local regions of any country, these various sources of emissions represent only one aspect of the potential air pollution problem; another important consideration involves local atmospheric conditions. The Soviet Union, like the United States, has many cities located in geographic regions that are climatically suited to pollution build-up. In general, these cities are situated in the more arid portions of the country, but Arctic regions, especially in winter, can suffer from stagnant air trapped below low-level inversions as well. *Inversions* are atmospheric conditions in which warmer air overlies cooler air near the earth's surface, creating a "lid" which traps and concentrates air pollutants. This phenomenon, first identified in Los Angeles, has also been observed in such Soviet cities as Alma-Ata and Yerevan. A particularly health-threatening situation seems to exist in the otherwise attractive city of Yerevan, the capital of Armenia. Here, metallurgical, chemical, and rubber factories, together with a regional power plant, produce an emission problem that is compounded by ideal conditions for the creation of photochemical smog (*Soviet Geography*, 1987, p. 608; Wolfson, 1988a). As automobiles become more common, and their emissions mix with those from industry and power plants, photochemical smog may become an increasing problem in many of the southern urban regions of the USSR.

To counter this threat, the USSR, like many other nations, is investigating alternative automotive fuels to serve as a substitute for gasoline. One of these is compressed natural gas (CNG), which is used to a small degree in fleet vehicles; the author had the opportunity to ride in such a vehicle in 1989 in Riga. However, *Pravda* reported in 1987 that only 30 percent of the nation's CNG production capacity was being utilized (*CDSP*, 1987, 38, p. 29).

Energy production

Among the most common stationary sources of atmospheric pollutants are fossil fuel power plants. The pollution potential will be especially severe if the fuel being burned is coal, which generally contains large amounts of ash and sulfur. The Soviet Union has constructed large coal-burning power plants in most parts of the country (Central Asia being an exception), with more under construction. These plants emit millions of tons of solid and gaseous pollutants annually. Containment of these gases and ash residuals is legally required, but effective abatement is enormously expensive, and the data in Table 2.2 indicate that the installation of equipment to capture these gases has largely stagnated during the 1980s. The growing problem of sulfur emissions in the context of acid rain will be reviewed in the next section.

One method to reduce the damage caused by coal-burning power plants is to convert them to either oil or, preferably, natural gas. Fuel oil produces fewer pollutants per kilowatt-hour (KWh) of generation than does coal, and gas is cleaner still. The use of coal in the Soviet Union has increased little, on average, since the mid-1970s. Per capita use of coal fell somewhat between the peak year of 1977 and 1985, but began increasing again in 1986, perhaps at least in part as a consequence of Chernobyl.

However, it is likely that economic factors, rather than environmental considerations, were the main cause of this. The cost of extracting Ukrainian coal has been rising, and its quality (in terms of average heat content) has been dropping, from 4,180 kilocalories per kilogram in 1965 to 3,840 in 1982 (Cooper, 1986). Despite this, and reflecting coal's great abundancy, the Soviet Union's long-term energy guidelines call for greater, not lesser, reliance on coal as an energy fuel in the 1990s ("Basic provisions," 1984, p. 13).

The substitution that the long-term energy plan *does* call for is to replace oil as a power plant fuel with natural gas. This has indeed been happening during the 1980s, with a large number of previous oil-burning plants undergoing this conversion. In the process, at least seven or eight coal-burning plants (in Perm, Zuyevka, Ladyshin, Shchekino, Yaroslavl, Kazan, and Ulyanovsk) have also been converted to gas (Cooper, 1986; Soviet Geography, 1990, 227–30). This latter step should have improved air quality in those cities.

The long-term energy plan of the USSR is designed to insure adequate future energy supplies, and to improve conservation and efficiency in the process. In terms of fuels, it stresses an increased reliance on nuclear energy and natural gas until the mid-1990s, and on nuclear energy and coal after that. It strongly stresses technical efficiency and conservation in energy use and development, especially in oil extraction. Here, there would appear to be considerable room for improvement; as a result of water being injected to increase the total oil obtained from a given pool, water represented over 50 percent of all the fluids raised from oil wells industry-wide, and as much as 81 percent in specific fields (Kelly *et al.*, 1986, p. 109).

The plan gives mention, but not emphasis, to fusion energy (see chapter 3) and to renewable energy sources ("Basic provisions," 1984, p. 15). Its year of publication (1984) predates both Chernobyl and Gorbachev's restructuring plans; the extent to which it may or may not have been revised in light of either of these considerations is unknown. It also gives brief mention to coal gasification, which (by use of additional energy) can transform dirty coal hydrocarbons into much cleaner methane. Readers interested in Soviet coal gasification research (and coal liquefaction as well) will find an extensive discussion of these topics in the work cited by Kelly *et al.* (1986).

It is of interest, particularly in light of the long-term energy program's emphasis on coal, that there is not one word in "Basic provisions" specifically dealing with atmospheric pollution, or with energy-related water consumption. The best that can be said is that the short-term emphasis on natural gas and the call for greater efficiency and conservation should have implicit positive effects on air quality.

In keeping with the program's intent, and Chernobyl notwithstanding, the Soviet Union has for many years been highly committed to the rapid construction of atomic powered electrical generating stations. About two dozen complexes are now in operation, mainly in the European part of the country, with many more under construction (see Table 3.2). The future of the USSR's nuclear program will be discussed in more detail in chapter 3. The other primary alternative to future fossil fuel power plants is renewable sources of energy, such as solar and wind. The extent of the Soviet Union's exploration of these options will be examined in chapter 4.

Acid rain

A contemporary atmospheric pollution problem of wide concern in both the United States and Europe is the phenomenon known as *acid rain*. The sources of this problem are sulfur and nitrogen gases emitted from industrial facilities and power plants, as well as nitric oxide from motor vehicles. When these combine with moisture in the air, they form nitric and sulfuric acids, which are carried to the earth's surface by precipitation. The ensuing run-off can increase the acidity of lakes to the point where fish life can no longer be sustained. Additionally, acid rain is probably responsible for widespread die-off of forests in Germany, and possibly in parts of the eastern United States as well. A complication factor is that these gaseous emissions frequently cross international borders; for example, from Germany into Scandinavia or from the United States into Canada.

The acid rain problem certainly exists in the Soviet Union, given the large number of sulfur emission sources found throughout the country. Compared to other environmental problems, however, relatively little has appeared in the popular Soviet press on this topic. One western specialist has suggested that "Soviet research on deforestation due to acid rain has apparently lagged behind research in severely affected Eastern European countries such as Czechoslovakia"

(Braden, 1988). The American journal *Environmental Science and Technology* published one of the earlier estimates for the USSR, reporting in its June 1984 issue that as much as 350,000 square miles (906,500 sq. km) of the Soviet Union may be affected by acid rain. Another study estimated that in 1980 the European part of the USSR alone produced 12.8 million tons of sulfur emissions, more than the UK, West Germany, France, Belgium, and Holland combined (Dovland, 1987). The latter was still true in 1984–85, although the USSR's sulfur emissions had declined in the interim by about 12 percent. The same source indicates, however, that on a *per capita* basis the USSR is a much smaller producer of nitrogen oxides than are the West European nations. At the beginning of this chapter, it was noted that sulfur dioxide emissions for the USSR in 1988 were given as 17.6 million metric tons.

The potential environmental effects of those sulfur emissions are sobering. Lakes in the northwestern part of the country are likely to suffer deterioration, and acidic precipitation could pose a threat to forests, wildlife habitat, and agricultural crops. *Time* magazine reported on Jan. 2, 1989 that the Baikalsk pulp mill near Lake Baikal alone had killed 86,000 fir trees. The effects of acid rain on forests are discussed further in chapter 7. Additionally, in aquatic environments, nitric acid has been shown to be convertible into nitrates, which can lead to the eutrophication of lakes and estuaries. Areas of the USSR with high sulfate levels included Moscow province, several areas in the Baltic republics, all heavy industry regions of the Ukraine, the central and southern Urals, and several heavy industry districts in Siberia and the Far East (Doklad, 1989, p. 25). Numerous technical articles have appeared in Soviet scientific journals which discuss the measured acidity and chemical composition of rainfall and run-off in specific locations, but generally without discussion of probable sources of the acidity.

But it is not difficult to suggest sources. Much of the coal mined in the Soviet Union has a relatively high sulfur content, and all fossil fuel power plants produce large amounts of oxides of nitrogen. All but one of the mining areas in the extensive Donbas coal basin, the nation's largest, produce coal with from 1.7 to 3.2 percent sulfur content. Also high (at c. 2.6 percent) is coal from the Volyn region and from the Moscow brown coal basin. Worst of all by a wide margin is coal from the Kizel mines which contains 6.1 to 8.4 percent sulfur; this coal ought not to be burned at all (Zalogin, 1979).

Other fields produce relatively better coal. Sulfur content in coal from the Chelyabinsk, Cheremkhovo, Karaganda, Ekibastuz, Kuznetsk Basin, and Kansk–Achinsk fields all averages under 1.2 percent. An environmental argument for the development of the new Kansk–Achinsk coal basin (see chapter 13) could be its low sulfur content (0.2–0.4 percent). However, even sulfur from low-content sources will wind up in the environment if it is not removed in the stacks. Even at Kansk–Achinsk, abatement equipment should be (and is being) used.

The USSR appears in the late 1980s to be viewing acid rain more seriously as a national (and international) problem. One contributing factor might be that at least 21 percent of the sulfur deposited in the USSR is estimated to originate outside the country (Dovland, 1987). Perhaps in partial response to this, the USSR in 1979 signed the United Nations Convention on Long Range Transboundary Air Pollution. To deal more specifically with the acid rain problem, in 1985 it signed the Helsinki protocol in which the signatory nations agreed to implement major reductions in sulfur oxide emissions. A 30 percent decrease was the recommended amount of reduction, and was the amount agreed to by the Soviet Union, although some nations said they would make cuts of over 60 percent. Far larger reductions than this may be needed, however, if desirable aquatic acid balances in many parts of Europe and North America are to be restored.

Ozone depletion

Gases that become transformed into acids are not the only "new" atmospheric problem to receive wide publicity in recent years. Another of at least equal concern involves the potential destruction of the protective stratospheric ozone layer (Cogan, 1988). Although ozone near the surface of the earth is an unstable and corrosive compound that is associated with photochemical smog, at higher elevations ozone is essential to life on the planet. High in the stratosphere, a concentration of ozone (a molecule of oxygen with three atoms) protects the earth's biosphere from the harmful effects of ultraviolet radiation.

The threat to the ozone layer lies primarily in the chemical action of certain escaped chemical contaminants called chlorofluorocarbons, or CFCs (Brasseur, 1987). Also involved are a related groups of bromine-containing compounds called halons, as well as smaller amounts of other gaseous emissions containing chlorine. The chlorine atoms in the CFCs react chemically in the stratosphere to destroy the protective ozone shield. A seasonal ozone "hole," representing an ozone loss of up to 50 percent, occurs each spring over Antarctica. In September of 1989, not only was this "hole" at close to record levels, but a smaller hole was observed over Australia as well. Smaller losses have been noted in high latitudes in winter near the North Pole, but some believe the potential for more significant Arctic losses is great (Kerr, 1989).

CFCs are used in a number of consumer and industrial applications, including the manufacture of refrigerants, air conditioners, aerosol spray cans, insulations, cleansers, solvents, and a number of other ubiquitous products. When these items reach the end of their useful life and are discarded, the CFCs are released to the atmosphere, where they react with, and destroy, ozone molecules. The loss of stratospheric ozone allows more ultraviolet radiation to reach the earth's surface which, among other things, can lead to an increase in the incidence of skin cancer in humans. It is believed that an increase of atmospheric CFCs can also contribute to an intensification of the "greenhouse effect," discussed below.

To combat this growing problem, an international conference was convened in Montreal in 1987. The "Montreal Protocol on Substances that Deplete the Ozone Layer" calls on all nations to reduce their use of CFCs by 50 percent by 1998 (developing nations were given a ten-year exemption). By 1989, thirty-two nations, including the Soviet Union, had signed it. At a subsequent conference in the spring of 1990, delegates from ninety-three countries, including all the European Community nations, the USSR and the United States, agreed to a complete elimination of CFCs by the end of 2005. Some countries that refused to sign the 1987 protocol, such as India and China, did agree to be a party to the stronger 1990 version, after the United States offered funding and technical assistance to third world nations to meet these goals. The DuPont corporation (and others) have agreed to manufacture viable substitutes. Although the USSR has said it intends to convert to non-CFC aerosols by 2005, at some conferences it had advocated a slower pace of CFC elimination. Given the evidence of an emerging problem over the north polar regions, it would seem logical for the US and the USSR to engage in a strong joint research program on this phenomenon.

The "greenhouse effect"

Another contemporary problem that involves atmospheric transformation in a possible adverse manner is termed the "greenhouse effect." As with the destruction of the ozone layer, this has the potential to produce major global changes. The "greenhouse effect" involves increased concentrations of the normally harmless gas carbon dioxide in the atmosphere. These concentrations have increased from around 280 parts per million (ppm) in 1800, to 315 ppm in 1960, to around 350 ppm in 1990. Other gases, such as the CFCs mentioned earlier, methane, and nitrous oxide also contribute to the warming. Soviet scientists have been leaders in studying changes in the atmosphere and their implications (e.g. Budyko, 1980; 1986).

The attribute of atmospheric carbon dioxide (and the other contributing gases) that is of concern is their ability to block outgoing long-wave radiation from the earth's surface, producing warming in the manner of the glass covering of a greenhouse. A doubling of the concentration of these gases could have the effect of warming the troposphere, eventually by perhaps 3 degrees centigrade or more (the exact amount is a matter of debate). In this regard, some scientists see cause for concern in the fact that six of the ten years in the 1980s were the six warmest on record, and 1988 was the warmest year in the last 130.

If significant long-term warming were to occur, it could produce a number of extremely adverse effects (Bolin et al., 1986). First, the warmer atmosphere would increase melting of polar ice caps. This melting polar ice could ultimately raise the level of the world ocean by as much as a meter or more, flooding at very high tides the low-lying parts of countries, and port cities, around the world. Holland, Bangladesh, and south Florida could be greatly endangered. Cities that could be affected in the Soviet Union include Leningrad, Riga, Tallinn, Murmanak, Odessa, Mariupol (Zhdanov), Sevastopol, Vladivostok, and many smaller towns. A raised ocean level could also produce increased coastal erosion.

Further, the warming would cause some mid-latitude regions of the earth to become more arid, including parts of interior and western North America and some southern portions of the USSR (Rosenberg, 1988; Revkin, 1988). On the other hand, some Soviet scientists also see possible beneficial results from this warming. For example, some models suggest that some northerly parts of the USSR might become warmer, and perhaps wetter as well, which could favor agricultural efforts in these areas (Budyko et al., 1986). However, it is important to note that not all computer models produce the same results, and that changes are expected to vary over both space and time (Revkin, 1988). Thus, the effects of global warming on Soviet agriculture could be mixed. In fact, some Soviet scientists have even questioned if any warming would occur as carbon dioxide increases (Borisenko and Kondrat'yev, 1984). It is generally felt that some areas of the world might benefit from potential atmospheric warming, while others might suffer greatly. More study of the problem is needed. Perhaps at least in part for this reason, the Soviet Union, Britain, Japan, and the United States in 1989 all resisted efforts to establish global limitations on carbon dioxide in the year 2000 at 1989 levels. It seems clear, though, that since the warming effect would be very hard to reverse if it occurs, much more study of the probable effects of atmospheric heating is needed.

The carbon dioxide that is partly producing the atmospheric warming results mainly from the burning of fossil fuels in vehicles and power plants. This forms another strong argument for converting from our present reliance on fossil fuels to the use of power sources that do not involve a combustion cycle. Both nuclear and renewable sources of energy take on added relevance in this context, and thus are the focus of chapters 3 and 4.

Administering air pollution: a summary

The Soviet Union acknowledges that it faces a number of serious air quality problems. Almost all industrial cities, including Moscow, suffer at least periodically from high atmospheric pollution readings; in some, the situation is admitted to be acute.

To address the problem, a national law was passed in 1980 setting air quality goals for the nation, and stipulating the types of regulations that should be enacted at the republic level ("Law . . . on air quality"). An elaboration on the numerous provisions of this law is provided by Ziegler (1987, pp. 88f). With regard to permitted pollutant levels, the general practice in the Soviet Union is to set very strict emission standards compared to other industrialized nations, which are based on optimal medical considerations (Derr *et al.*, 1981; Table 2.4). These standards, termed "Maximum Permissible Concentrations" (MPCs), establish admirable goals, but they frequently may be unattainable in practice. They have been supplemented in the 1980s by maximum allowable emission norms, which permit much easier control at the source of the emission. However, it again should be noted that enforcement difficulties exist which are caused by recurring budgetary and priority constraints.

Administrative responsibility for air quality has historically been split among a number of ministries and committees (Ziegler, 1987). In

Table 2.4 *Some comparative workplace ambient air quality standards (Milligrams per cu. m.)*

Pollutant	USSR	USA
benzene	5	30
carbon monoxide	20	55
lead	0.01	0.05
mercury	0.01	0.05
nitrogen oxides	5	9
ozone	0.01	0.02
sulfur dioxide	10	13
hydrocarbons	300	500

Note: These standards generally represent an average figure over a proscribed number of hours; the number of hours used is variable both by pollutant and by country. See source for specific details
Source: Derr, *et al.*, 1981, p. 32

the 1970s, monitoring of air quality was made the responsibility of the State Committee on Hydrometeorology (Gidromet), but it was never clear how much enforcement capability this agency was given. Perhaps because of this uncertainty, the new State Committee on Environmental Protection (Goskompriroda) which was described in the opening chapter has been created. In 1988 it was operating a network of 1,155 air quality monitoring stations in 537 Soviet cities (*Okhrana*, 1989, p. 33). It is to be hoped that under *perestroika* it will have the authority to implement stronger corrective actions but, as subsequent chapters will discuss, there is reason to question whether this in fact will be the case. Research on improving air quality is carried out at a number of technical institutes, and is reported upon in the journal *Problemy kontrolya i zashchita atmosfery ot zagryazneniya* (Problems of control and protecting the atmosphere from pollution).

In any country, improving air quality requires a systems approach to the problem. In both capitalist and socialist nations, decisions are made throughout the economy that have significant implications for increased energy demand, with little if any thought given to the pollution that will be generated from power plants as a consequence. For example, a study conducted at the University of Glasgow raises the interesting point that substituting electric furnaces in the steel industry for older, "dirtier" processes will not necessarily produce less net national air pollution, if brown coal or low grade oil is burned in the electricity-producing power plants.

In addition to the quality of local air basins, long-range alterations to the chemical make-up of the world's atmosphere as a whole have become a compelling issue. For some of these issues, such as the destruction of the protective ozone shield, a consensus on the need for immediate action seems to be emerging. For others, such as the "greenhouse effect," considerable differences of opinion still exist, with at least some Soviet and American scientists not in full agreement on the implications of global warming, or even on its inevitability. Yet there is agreement that these problems are of great concern, and represent an opportunity – indeed, many might say an imperative – for cooperative efforts to avert major environmental deterioration of our shared global biosphere in the future.

Appendix 2.1 Discharge of air pollutants from stationary sources for selected USSR cities, 1987 (emissions data in 1,000s of tons)

City	Population (1,000s)	Per capita emissions[a]	Total	= Solids	+ Gases and liquids	Gases and liquids: Sulfur dioxide	Nitrogen oxides	Carbon monoxide	Other gases and liquids
Alma-Ata	1,108	43.4	48.1	10.7	37.4	15.7	3.8	15.2	2.7
Arkhangelsk	416	200.0	83.2	22.7	60.5	38.4	7.5	12.2	2.4
Ashkhabad	382	49.5	18.9	3.6	15.3	.4	.2	14.1	.6
Baku	1,741	271.2	472.2	182.3	289.9	20.3	16.5	59.6	193.5
Bratsk	249	694.8	173.0	40.7	132.3	22.0	4.7	100.5	5.1
Chelyabinsk	1,119	375.4	420.1	109.6	310.5	60.7	29.4	212.4	8.0
Dzhambul	315	370.2	116.6	41.0	75.6	51.7	13.9	3.5	6.5
Donetsk	1,090	178.1	194.1	23.8	170.3	33.7	7.2	120.7	8.7
Dushanbe	582	64.4	37.5	14.2	23.3	8.3	4.3	9.5	1.2
Frunze	632	137.3	86.8	23.9	62.9	42.1	7.9	10.6	2.3
Irkutsk	609	146.8	89.4	30.2	59.2	26.4	8.1	23.4	1.3
Kiev	2,544	36.9	93.8	15.0	78.8	39.1	22.1	5.9	11.7
Kemerovo	520	258.8	134.6	41.0	93.6	23.9	29.7	27.7	12.3
Kishinev	663	63.2	41.9	4.7	37.2	18.8	5.4	8.7	4.3
Krasnoyarsk	899	327.3	294.2	85.9	208.3	40.2	12.6	139.6	15.9
Leningrad	4,948	52.7	260.9	50.7	210.2	85.0	45.5	47.3	32.4
Mariupol' (Zhdanov)	529	1,485.4	785.8	115.8	670.0	55.0	30.1	577.5	7.4
Magnitogorsk	430	2,026.5	871.4	180.8	690.6	81.7	34.4	561.7	12.8
Minsk	1,543	72.2	111.4	10.9	100.5	29.5	17.3	39.4	14.3
Moscow	8,815	41.6	367.1	36.2	330.9	113.8	117.8	25.6	73.7
Mogilev	359	375.2	134.7	6.8	127.9	70.6	7.5	37.8	12.0
Novokuznetsk	589	1,515.8	892.8	153.7	739.1	90.9	36.7	599.7	11.8
Odessa	1,141	93.6	106.8	23.3	83.5	20.1	6.7	29.9	26.8
Riga	900	45.6	41.0	10.2	30.8	9.9	2.8	11.5	6.6
Tallinn	478	86.2	41.2	7.3	33.9	19.7	4.2	7.3	2.7
Tashkent	2,124	23.9	50.7	15.9	34.8	4.0	4.9	18.4	7.5
Tbilisi	1,194	34.6	41.3	8.1	33.2	4.0	3.4	21.1	4.7

Ust'-Kamenogorsk	321	450.5	144.6	26.2	118.4	73.3	8.1	36.0	1.0
Ufa	1,092	318.2	348.2	9.2	339.0	101.8	27.1	42.2	167.9
Volgograd	988	283.4	280.0	68.8	211.2	46.4	19.6	58.8	86.4
Vilnius	566	66.1	37.4	2.2	35.2	22.0	4.9	5.8	2.5
Yerevan	1,168	62.5	73.0	14.3	58.7	19.6	11.7	12.0	15.4
Zaporozh'ye	875	328.1	287.1	87.6	199.5	27.4	14.6	145.2	12.3
Sum for cities	40,929	175.4	7,179.8	1,477.3	5,702.5	1,316.4	570.6	3,040.8	774.7
USSR reported total	281,689	227.6	64,100.0	15,600.0	48,500.0	18,600.0	4,500.0	15,500.0	9,900.0
Selected cities as a percent of USSR reported total	14.5		11.2	9.5	11.8	7.1	12.7	19.6	7.8

Note: [a]Equals total stationary emissions (third column) divided by population

Sources: Narodnoye khozyaystvo SSSR v 1987g., p. 573; per capita data and national percentages as calculated by Sagers in *Soviet Geography,* June 1989, pp. 514–15

3   The Chernobyl legacy: whither nuclear energy?

Among contemporary environmental issues, few are more controversial than the role and long-term ramifications of nuclear energy. Following its initial military use to draw World War II to a close, it was quickly transformed into "the peaceful atom." In some circles in both the United States and the USSR, it almost took on the aura of an energy panacea. This aura was eventually dimmed by Three Mile Island, and was seemingly eradicated in 1986 by the Chernobyl accident. Or was it? Today, equally serious concerns about atmospheric warming, already discussed in chapter 2, are forcing all countries to re-evaluate the future of the nuclear option. Optimal environmental management will soon necessitate some hard decisions.

The Soviet Union lays claim to originating the production of electricity from atomic energy. Soviet scientists placed the first 5 megawatt (MW) commercial nuclear reactor into operation at Obninsk, near Moscow, in June of 1954. Since then, the USSR has enthusiastically pursued research in all aspects of nuclear development: fission, fusion and breeder reactors, as well as non-electrical applications such as atomic icebreakers and underground oil and gas stimulation. Soviet research on commercial nuclear reactors is available in an English translation of the journal *Soviet Atomic Energy*.

The Soviet commercial reactor program

As of January 1, 1990, the Soviet Union had in place approximately 38,000 megawatts (MW) of nuclear generating capacity, two-thirds of which was installed in the 1980s. This places the USSR first in the world in terms of nuclear capacity constructed during this decade. It also raised the proportion of nuclear energy in the nation's electrical grid to over 11 percent of the total (Table 3.1). This percentage is expected to increase slightly as new atomic facilities come on line during the 1990s.

Table 3.1   *Growth of the Soviet commercial nuclear power industry*

End of year:	Number of commercial reactors:	Installed MW (a)	Percent of all electrical generation
1965	2	310	< 1
1970	4	875	1
1975	12	4,845	2
1980	22	12,700	6
1985	38	27,520	10
1987	45	35,020	11
1990 (est.)	50	c.40,000	c.12
1995 (est.)	60	c.50,000	c.12

Note: a Commercial plants only, does not include a 600 MW plant in the Urals used for weapons production
Sources: *Soviet Geography*, April 1986; Petrosyants, 1987; Sagers, 1988

The determined pace of the Soviet nuclear effort can be seen in the fact that installed capacity doubled in each five-year period between 1965 and 1985. Although this pace could not be maintained during the 1985–90 five-year plan due both to the effects of Chernobyl and a finite capital funding capability, it is likely that about another 13,000 megawatts of capacity will have been added during this period. This would mean that in 1990 nuclear power output will represent about 11–12 percent of the total national electrical output. This is considerably under the 20 percent (based on 41,000 MW being added during 1986–90) that was originally envisioned in the plan (Ryzhkov, 1986). An even smaller increase is now envisioned for the 13th five-year plan, 1991–95 (Table 3.1).

Within the Soviet Union, nuclear power stations are found almost exclusively in the western, or European portion, of the country. This is where most of the population and industry is located, but it is also a region that today has limited developable resources of hydrocarbon fuels and hydroelectric power. West of the Urals economic region the major oil deposits (Caucasus and Volga-Ural) have all peaked, Donbass coal is now deep below the surface and expensive to extract, and almost all significant hydroelectric sites have already been developed. Also, the USSR has power delivery commitments to Eastern Europe. Thus, the expectation has been (at least prior to Chernobyl) that most new electrical power west of the Urals will come from nuclear energy, despite its high capital cost and radiation problems. In

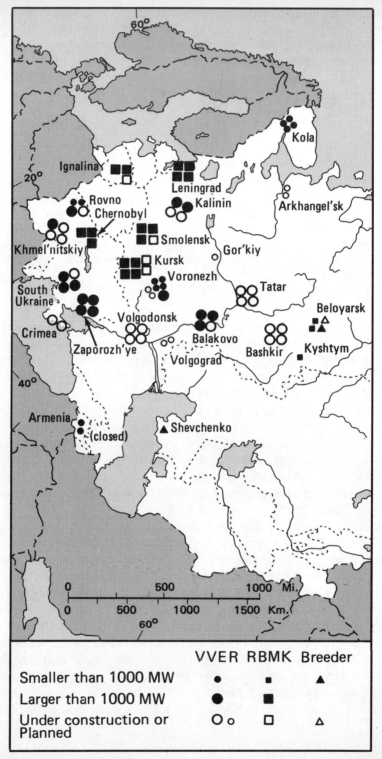

Figure 3.1 Soviet nuclear reactor sites

addition to fossil fuel scarcity, air pollution problems, and the threat of global warming have played a role in this decision.

In Siberia (east of the Urals economic region), with its enormous resources of coal, oil, natural gas, and hydropower, nuclear energy development is generally unnecessary. At present, the only exception is the very small generating station (48 MW) at Bilibino, located in a remote corner of the Chukchi tundra. Although the 13th (1991–95) five-year plan envisions new units in the Urals region in the Bashkir autonomous republic, at a "South Urals" site (exact location unspecified), and possibly a new breeder at Beloyarsk, no new commercial nuclear facilities are planned for any location east of the Urals foothills (Figure 3.1).

Although many different designs of nuclear reactors exist, almost all existing reactors in the USSR are of one of two types. About half are the pressurized water type, which use water for cooling and moderating (in the USSR these are identified by the Russian acronym VVER). The remaining half, including the Chernobyl units, are graphite moderated (with the Russian acronym RBMK); they are also sometimes referred to as "channel" or "pressure tube" reactors. The difference between the two designs are well described in such works as those by Kelly *et al.* 1986, by Dienes and Shabad, 1979, and in the April 1983 issue of *Soviet Atomic Energy*. The RBMK (graphite moderated) reactor design is described in detail in Gittus *et al.* The graphite moderated design is little used outside the USSR, being found in the United States, for example, only at older weapons-grade material production sites. Gas cooled reactors, although a common design in Europe, have not been put into commercial use in either the Soviet Union or the United States.

Most of the new nuclear reactor units built in the USSR in the late 1970s and early 1980s were of the graphite (RBMK) type, and were located at such major complexes as Leningrad, Kursk, Chernobyl, and Smolensk. However, in the late 1980s most of the newer sites were utilizing pressurized water (VVER) reactors, e.g., at the Rovno, South Ukraine, Kalinin, Balakovo, Khmelnitskiy, and Zaporozhye complexes, as well as at the earlier stations near Yerevan and Voronezh (Table 3.2). The Khmelnitskiy reactors are of particular interest to Europe, since by CMEA agreement half of their output will be exported to Poland, Hungary, and Czechoslovakia (see Figure 6.1). The first unit at Khmelnitskiy went on line in 1987, three years late. The decision to re-emphasize VVER reactors was made well prior to the Chernobyl accident, though whether the main reason for this was economics or safety is not clear.

Table 3.2 *Commercial nuclear energy sites in the USSR*

Name of complex	Site location [a]	Reactor type	Reactor size [b]	First unit in operation
Arkhangelsk	Arkhangelsk [h]	VVER	500	199?
Armenia	Metsamor	RBMK [c]	440	1976 [i]
Balakovo	Balakovo	VVER [d]	1,000	1985
Bashkir	Agidel	VVER	1,000	199?
Beloyarsk	Zarechnyy	RBMK [e]	(f)	1964
Chernobyl	Pripyat	RBMK	1,000	1977 [i]
Gorkiy	Gorkiy [h]	VVER	500	1989
Ignalina	Snieckus	RBMK	1,500	1983 [i]
Kalinin	Udomlya	VVER	1,000	1984
Khmelnitskiy	Neteshin	VVER	1,000	1987
Kola	Polyarnye Zori	VVER	440	1973
Kursk	Kurchatov	RBMK	1,000	1976
Leningrad	Sosnovyy Bor	RBMK	1,000	1973
Novovoronezh	Novovoronezh	VVER	1,000; 440	1964
Rostov	Volgodonsk	VVER	1,000	199? [i]
Rovno	Kuznetsovsk	VVER	1,000; 440	1980
Shevchenko	Shevchenko	Breeder	150	1973
Smolensk	Desnogorsk	RBMK	1,000	1982
South Ukraine	Konstantinovka	VVER	1,000	1982 [i]
Tatar	Kamsk. Polyany	VVER	1,000	199?
Volgograd	Volgograd [g]	VVER	500	199?
Voronezh	Voronezh [h]	VVER	500	199?
Zaporozhye	Energodar	VVER	1,000	1984

Notes: [a] See Figure 3.1. Not included are sites near Minsk, Kersk (Crimea), Odessa, Kharkov, and Krasnodar where planned reactors have apparently been cancelled, and a "South Urals" complex where neither the site nor specifics concerning the reactors have been disclosed
[b] In megawatts (MW)
[c] RBMK = Graphite moderated reactors
[d] VVER = Pressurized water reactors
[e] There are two RBMK reactors and one breeder reactor at Beloyarsk
[f] The RBMK reactors are 100 and 200 MW in size; the breeder is 600 MW
[g] Reactors provide both electricity and district heating
[h] Reactor is used to provide district heating only
[i] One or more units at this location shut down or cancelled during 12th five-year plan (1986–90)
Source: Soviet Geography, Apr. 1988, pp. 448–54, Apr. 1989, pp. 338–45, Apr. 1990, pp. 310–13

The Soviet Union, unlike the United States, has standardized the design of its nuclear power plants. Not including early prototypes and the breeder reactors, four main designs have been adopted. The smallest is the 440 MW pressurized water unit (termed the VVER-440); this reactor is exportable and has been (or will be) constructed at several sites in East Europe, Finland, and Cuba. The 1,000 MW VVER reactor was first proven out at Novovoronezh, and is now being mass-produced for use at the numerous locations mentioned earlier. The third and fourth designs are 1,000 and 1,500 MW graphite moderated units of the RBMK type. At present, the giant RBMK-1,500 units (the largest in the world) are being used only at the Ignalina site in Lithuania. The Chernobyl accident will probably terminate earlier plans to build them elsewhere in the future. Reportedly, no new atomic power stations, other than those now in operation or under construction, will utilize RBMK reactors (*Soviet Geography*, 30, 1978, p. 343).

In order to supply the number of reactor units now being called for, a huge new factory (termed "Atommash," an acronym for "atomic machinery") to mass-produce the large VVER units has been built at Volgodonsk, a fast-growing city upstream from Rostov on the Don River. It was designed to turn out three to four 1,000 MW reactor units a year, but actual annual production to date has been less than half this. The reasons range from serious foundation problems at the plant caused by inadequate geological studies, to a variety of management and operational problems that have raised questions about the quality of Atommash's products (*Pravda*, July 20, 1983, p. 2; Marples, 1986, p. 103f). The housing for the initial reactor unit, for use at the South Ukraine complex, was completed at Atommash early in 1981.

Breeder reactors are highly sophisticated devices that are able to produce more fissionable fuel than they consume. Since (a) their cost is very high, (b) they are inherently risky, and (c) one breeder can supply many conventional reactors with fuel, few exist at present (and none at all in the United States).

The Soviet Union first built an experimental breeder reactor in 1959, and in 1972 completed construction on its first commercial breeder (the BN-350) at Shevchenko on the east side of the Caspian Sea. This facility has the dual purpose of providing electrical output and powering a nearby desalinization plant. A larger breeder reactor (designated the BN-600) exists at Beloyarsk, adjacent to two small RBMK units at the same site. It was completed in 1980, seven years later than scheduled (Kelly *et al.*, 1986, p. 65). There are reports that a still larger breeder of 800 MW may be built at or near Beloyarsk, but its construction status at present is unclear (*Soviet Geography*, 30, p. 344). With its two existing units, the USSR joins the United Kingdom and France as the major breeder reactor nations in the world. Information on Soviet breeder reactors can be found in the April 1983 issue of *Soviet Atomic Energy*.

All existing commercial reactors are of the fission type, which split uranium atoms, thereby producing heat energy. Fusion reactors, which produce even more energy by combining very light weight atomic nuclei, are at present still in the experimental stage. They have several attractive features, however; they cannot explode, they do not produce high level radioactive wastes, and their input fuels exist in abundance. Thus, fusion reactors are of great interest in the technologically advanced countries.

The USSR has long carried out some of the most sophisticated fusion research in the world, and indeed pioneered the "tokamak" fusion concept which is now the most widely used research design. The Soviet Union presently has several experimental fusion devices in operation, employing a wide variety of different designs (Kelly *et al.*, 1986, pp. 73ff). The USSR's long-range energy plan sees them as "one of the most likely areas for the creation of a virtually inexhaustible energy supply source" ("Basic provisions," 1988). Due to their great complexity, however, commercial fusion reactors are probably at least twenty years into the future. Among the primary Soviet fusion research centers are the Kurchatov Institute in Moscow, and related facilities at Kharkov, Leningrad, Novosibirsk, and Sukhumi; joint research programs exist with the United States and other countries.

The question of safety

The Soviet Union, like the United States, has tried to popularize the idea that nuclear power plants are a safe and reliable way to generate electrical energy. A leading journal of the Soviet nuclear power industry once stated about its VVER reactors, as a typical example in the pre-Three Mile Island (and pre-*glasnost'*) era, that: "there is every reason to consider [these] plants no more dangerous than conventional power plants. Radiation injury to the population is practically impossible, and any presumable emergency situation in nuclear power stations with water cooled, water moderated reactors cannot be of a catastrophic nature" (Ostachenko *et al.*, 1971).

Despite official assurances, it was clear from press reports that at least a segment of Soviet society remained unconvinced about nuclear safety. Perhaps not coincidentally, shortly after the Three Mile Island accident in 1979, official publications began to suggest a need to consider carefully the many ramifications of a large-scale nuclear program (Dollezhal and Koryakin, 1980). At about the same time, popular concern in the Soviet Union seems to have increased, as evidenced by occasional artic'es in the press designed to dispel public anxieties (e.g., *Trud*, Sept. 26, 1982; *Sots. industriya*, Sept. 21, 1983; *Pravda*, Jan. 30, 1984).

In fact, official Soviet reaction to the Three Mile Island episode was interesting, ambiguous, and perhaps revealing. A fortnight after the accident (April 11, 1979) the Soviet press carried a strident article that characterized Three Mile Island as "a serious, major accident, one that threatened at any moment to turn into a catastrophe, even a terrible tragedy" (*Lit. gazeta*, p. 9). Yet on the very same day, another article complained that: "(T)he Western press's treatment of the atomic reactor accident at Harrisburg, in which essentially minor unfavourable consequences were depicted in an extremely exaggerated form, was a continuation of the campaign against atomic power" (*Izvestiya*, p. 2).

Nuclear plant safety in the USSR is governed by radiation standards that were initially drawn up by the Ministry of Health in 1969, and revised in 1973 and 1976. They are enforced by the State Sanitary Inspectorate. An All-USSR State Committee for Atomic Power Safety was created in the summer of 1983. Radiation from properly operating Soviet atomic power stations, like those elsewhere, is less than normal natural background radiation.

Each Soviet nuclear power plant must maintain a radiation safety department that has responsibility for meeting these standards, for monitoring ambient radiation levels both in the plant and in the surrounding environment, and for ensuring the safety of plant personnel. Both standards and radiation monitoring results at Soviet nuclear power plants are available in English translation (*Soviet Atomic Energy*, Apr. 1983, pp. 290ff). Interestingly, even before the accident, worker exposure to radiation was higher (though within limits) at Chernobyl than at other stations, at least in early years (Table 3.3).

However, a delegation of nuclear scientists from England who visited Soviet RBMK reactors in 1975 came away very concerned with plant safety, noting nine features of these reactors that made them uneasy. According to a later British report, this type of design would not even have been considered in the UK (Gittus *et al.*, 1987). Both the basic design of RBMK reactors and plant operating procedures have

Table 3.3   *Average annual personal radiation dose, rem*

Atomic power station	1977	1978	1979
Novovoronezh	0.78	0.61	0.60
Kola	0.84	0.59	0.68
Armenia	0.33	0.14	0.62
Chernobyl		1.20	1.00
Kursk	0.23	0.33	0.38

Source: Soviet Atomic Energy, Apr. 1983, p. 292

been re-evaluated, and hopefully corrected, in the wake of the disaster at Chernobyl.

Chernobyl and its predecessors

Prior to Chernobyl, accidents involving commercial nuclear facilities had occurred in the USSR, but were never immediately publicized. One such incident, believed to be a steam line explosion, took place in 1973 at the Shevchenko breeder reactor. Another, in 1978 at the unit 2 graphite reactor at Beloyarsk, involved a major fire in the control room that made shutting down the crippled reactor difficult. A *glasnost'*-era article stated that only a miracle averted a meltdown, and that had this accident not been covered up for almost a decade, Chernobyl might have been prevented (*Sots. industriya*, Oct. 21, 1988). A very serious early accident, involving radiation releases from a non-commercial nuclear processing or storage site, will be discussed in chapter 6. As with nuclear plants elsewhere, numerous "unplanned shutdowns" occur throughout the system annually, including one at Kursk in 1989 involving the release of a small amount of radiation.

In addition to accidents, complaints about the quality of work at construction sites, and of nuclear plant components, have appeared in the Soviet press. Two sites where workmanship has been frequently criticized are at Bilibino and Ignalina (Marples, 1986). The 1983 problems at the Atommash factory were noted earlier.

The 1986 accident at Chernobyl's unit 4, of course, could not be concealed. Many locales in several West European countries, as well as in the Soviet Union, were contaminated with high-level radiation, causing serious problems for drinking water supplies, food pro-duction, and livestock (Wynne, 1989). One estimate suggests that Chernobyl released about 10,000 times as much radioactivity as the next worst nuclear plant accident, that at Windscale (Sellafield) in England in 1957 (Hohenemser and Renn, 1988). It may have been the most expensive accident in the history of the world, with early clean-up estimates running as high as 13 billion rubles, costs that required diverting vast amounts of resources from other sectors of the economy. By 1990, total clean-up costs were being estimated as high as 40 billion rubles. It is widely suggested, even in the Soviet Union, that the full consequences of Chernobyl are still being understated today; this question was the subject of a long article in *Izvestiya* on March 26, 1990.

A detailed recounting of the Chernobyl accident is unnecessary here, it has been amply reported elsewhere (Abagyan *et al.*, 1986; Shabad, 1986; Marples, 1986; Gittus *et al.*, 1987; Hohenemser and Renn, 1988). A brief review of the main environmental consequences of the accident will suffice.

The immediate human and environmental effects were staggering. Thirty-one fatalities were reported, though in 1989 *Moscow News* suggested a more inclusive figure would be around 250, and in 1990 various sources were suggesting over 300. To minimize human exposure to residual radiation, all persons within a 30 km radius from the plant site were evacuated. Up to 135,000 people were initially relocated, 50,000 of them from the cities of Pripyat and Chernobyl (Figure 3.2). Over 24,000 of the evacuees resided within Belorussia, though a more recent estimate suggests 106,000 Belorussians will need to be resettled (*Pravda*, July 30, 1989).

Pripyat and Chernobyl may become permanent ghost towns; possibly they will be razed. One Soviet specialist commented that "today ... the radiation situation in [the town of] Chernobyl is such that full-fledged human life cannot be resumed there for decades" (Levada). A new town for the power plant workers, named Slavutich, is being built just outside the 30 km circle. It is possible the radiation levels are above normal in this area, as well. The area in which radiation contamination exceeded maximum permissible norms was reported in 1989 to include over 1500 sq. km in the Ukraine, about 2,000 sq. km in the Russian republic, and some 7,000 sq. km in Belorussia (*Pravda*, Mar. 20, 1989). *Pravda* also noted that about 250,000 people live on land so contaminated that their food must be shipped in. In April of 1990, the Soviet government allocated an additional $26 billion for assisting the 3 to 4 million persons who reside in areas suffering some degree of contamination.

On April 17, 1990 *Pravda* published more detailed information on

Figure 3.2   Chernobyl and vicinity *Source: Atlas SSSR, 1983;* Doklad ... 1989, p. 44

Table 3.4   *Areas contaminated with cesium 137*

Republic	Total area contaminated (sq. km)			Of which, outside evacuation zones		
	5–15[a]	15–40[a]	40 +[a]	5–15[a]	15–40[a]	40 +[a]
Belorussia	10,160	4,210	2,150	9,830	3,640	1,160
Ukraine	1,960	820	640	540	350	200
Russia (RSFSR)	5,760	2,060	310	5,760	2,060	310
TOTALS	17,880	7,090	3,100	16,130	6,050	1,670

Note: [a]Curies of Ce-137 per square kilometer
Source: Izrael, 1990

the extent of the area contaminated by the Chernobyl accident; these data are presented in Table 3.4 (Izrael, 1990). They confirm that the most seriously affected republic was Belorussia, not the Ukraine, and that 1,670 sq. km with the highest measured levels of Cesium 137 still contain unevacuated populations.

In order to keep contaminated dust from blowing, techniques such as sprinkling exposed earth, asphalting dirt roads, and even spraying thin plastic films were used. Hay and other cut crops had to be collected and disposed of; farm animals and household pets had to be rounded up and relocated. Towns and villages approved for rehabitation had to be thoroughly decontaminated. The destroyed reactor itself had to be completely entombed in reinforced concrete at great expense.

There was wide concern for the safety of food products. Fresh milk could not be drunk by Belorussian and Lithuanian children, fishing was banned in the Dnieper, Pripyat, and Sozh rivers, street vendors selling food products disappeared from Kiev, mushroom gathering was prohibited, and fresh vegetables disappeared from many markets. There have been occasional reports, however, that "tainted" meat from the area was being sold in other parts of the USSR. It was not known when agricultural land within the 30 km zone (and in some spots outside it) could safely be used again for food crops, if ever. Possibly industrial crops, such as flax, or non-commercial crops that absorb radionuclides, such as lupine, might be grown in less contaminated areas.

Water contamination was of critical concern: 20 km of dikes had to be constructed to prevent run-off from entering the Pripyat River, which flows into the Kiev Reservoir from which the city of Kiev draws about half its drinking water. To circumvent this problem, a new pumping station was built on the Desna river to make this water source accessible to Kiev, and 58 new deep artesian wells were dug in the city (Shabad, 1986). Thousands of wells in the two republics had to be sealed or pumped out due to contamination. It is widely believed, however, that at least some radioactivity must have inevitably made its way to the Kiev Reservoir.

It is almost certain that the full human and environmental costs of the accident are not yet known, and will not be for some time. The initial lack of candor about the accident, especially in the Belorussian republic, is now acknowledged and lamented even by the Soviet government (*CDSP*, Nov. 29, 1989). Long-term monitoring of the biosphere in the vicinity of Chernobyl will be essential. The effects on waterfowl and other birds that migrate through the region are unknown, but in 1989 *Moscow News* reported that sixty-four deformed pigs and calves have been born at a nearby farm since the accident. Certainly the most serious of the delayed effects is the possibility of future cancer occurrences, and the same issue of *Moscow News* noted marked increases in cancer rates and thyroid gland problems in an unspecified area 50 km from Chernobyl. Some scientists feel, however, that the chances of induced cancers outside the USSR are slight (Gittus, *et al.*, 1987; Anspaugh *et al.*, 1988).

Although there has been detailed reporting on the effects of the disaster within affected areas of the Ukraine, until 1989 there was surprisingly little reported from nearby Belorussia, despite a known high level of fall-out there (Figures 3.3 and 3.4). Indeed, an estimated 70 percent of all Chernobyl fallout in the USSR landed on Belorussia. Then, early in 1989, almost three years after the accident, it was decided to evacuate twenty additional villages in Gomel and Mogilev provinces, due to persistently high levels of cesium-137 in the area. Public concern over radiation in the republic was a factor in cancelling the proposed nuclear power station near the capital city of Minsk. In addition, in August of 1989 Tass announced that 3,000 people in 12 villages would also be evacuated from the area in the Russian republic (Bryansk province) that is closest to Chernobyl.

Finally, it is sobering to consider what could have happened if the events leading to the accident had occurred at the Leningrad atomic station at Sosnovyy Bor, which also houses four RBMK graphite reactors (see Figure 5.2). This plant is located only 60 km upwind (west) from Leningrad; a Chernobyl-type accident there might have seriously contaminated millions.

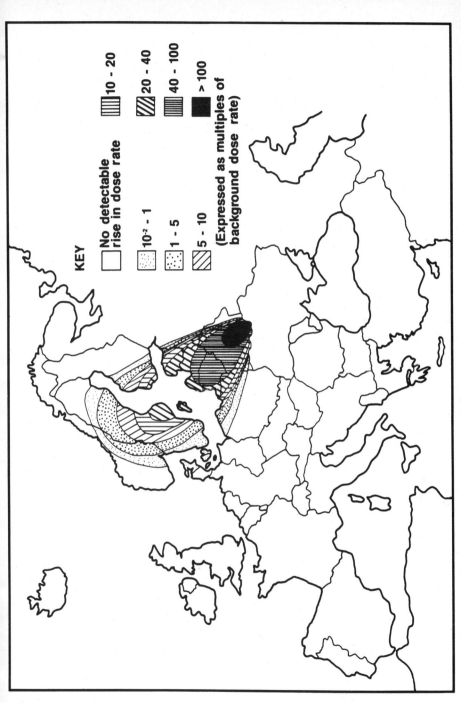

KEY

☐ No detectable rise in dose rate	▥ 10 - 20
⦂ 10⁻² - 1	▨ 20 - 40
⦂ 1 - 5	▤ 40 - 100
▧ 5 - 10	■ > 100

(Expressed as multiples of background dose rate)

Figure 3.3   Extent of radiation dispersal, April 29, 1986 *Source:* Gittus *et al.*, 1989, reprinted by permission of the United Kingdom Atomic Energy Authority. Copyright UKAEA

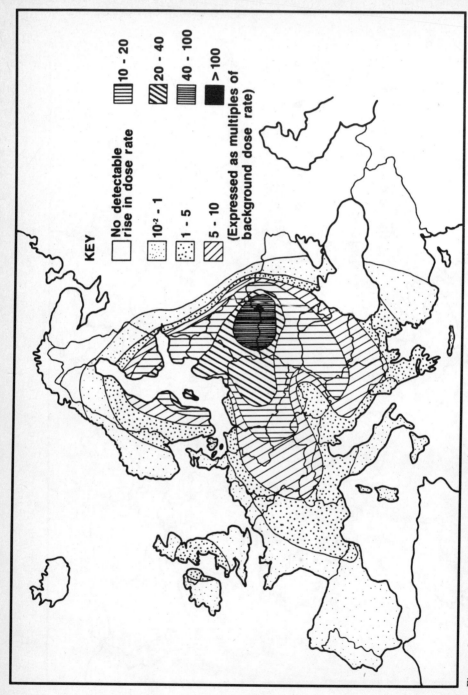

KEY

No detectable
rise in dose rate

10^{-2} - 1

1 - 5

5 - 10

10 - 20

20 - 40

40 - 100

>100

(Expressed as multiples of
background dose rate)

Figure 3.4 Extent of radiation dispersal, May 3, 1986 *Source: Gittus et al.*, reprinted by permission of the United Kingdom Atomic Energy Commission. Copyright UKAEA

The aftermath of Chernobyl

The Chernobyl accident necessitated a critical review of a number of aspects of the nuclear power program in the USSR, not just the operational procedures at the RBMK (graphite) reactors (Potter, 1990). Among these are plant safety at *all* Soviet reactors and processing facilities, regardless of design, the question of constructing containment domes, the eventual decommissioning of plants, the wisdom of the new "heat-and-power" nuclear plants, and of course public perceptions and concerns. Most importantly, the environmental consequences of Chernobyl will have to be monitored well into the twenty-first century.

Some of the above considerations are deserving of a little elaboration. Most Soviet nuclear reactors are built without the type of containment domes used on reactors in most other countries. The first Soviet reactor to utilize a containment structure was the 1,000 MW pressurized water reactor at Novovoronezh. A 1987 article in the government newspaper seemed to indicate that all of the new VVER reactors of this type would have containment domes; however, the article did not mention their use in connection with RBMK graphite reactors (*Izvestiya*, Dec. 9, p. 3). Soviet nuclear engineers appear to still consider the RBMK reactors inherently safer than the pressurized water ones, *assuming* no catastrophic malfeasance takes place as was the case at Chernobyl. Nevertheless, the RBMK design will apparently be abandoned in the USSR in the future.

A basic concern is the safe disposal and long-term storage of high-level nuclear wastes. By the year 2000, the Soviet Union plans to have installed more nuclear power plant capacity than the United States, with additional large amounts of high-level wastes being generated by military programs. Soviet practice is to store these wastes initially for up to three years at the reactor site (ten years for VVER-440 reactors). Then they are transported to reprocessing sites, usually by train but occasionally by boat or truck (*Soviet Atomic Energy*, Apr. 1983, pp. 309ff).

The Soviet Union buries high-level wastes from such plants in deep, geologically stable formations, possibly after vitrification. The exact location of the burial sites has never been revealed. The Chernobyl clean-up has produced enormous additional amounts of high-level radioactive earth and debris, possibly in volumes not amenable to deep burial. These contaminated by-products must also be stored in stainless steel tanks, although vitrification may be used (Marples,

1986). As for wastes with lower levels of radiation, at least some of these are solidified and stored in a retrievable manner below the surface (*Pravda*, May 15, 1986, p. 6).

Water is another aspect of nuclear power plant operation that requires attention. Power plants of all types may collectively use as much as 100 cu. km of water annually, with 2 cu. km of this being lost to evaporation in European USSR alone. These requirements could double by the year 2000 (Dollezhal and Koryakin, 1980). Some of the water quality problems described in chapter 5 have probably been caused at least in part by the huge demands of power plants. Deterioration of the lakes near the reactor complex at Ignalina has been especially mentioned in this regard.

There is also concern for safety at the front end of the nuclear fuel cycle. The Soviet uranium mining, processing, enriching, and fuel fabrication industries have only rarely been reported upon, due to the secrecy that surrounds these operations (Dienes and Shabad, 1979;) *Soviet Atomic Energy*, 1983, pp. 301ff). Yet experience in other countries has shown that the radiation danger in these early phases of the nuclear fuel cycle can easily be underestimated, and the probable events at Kyshtym (see chapter 6) raise questions concerning other aspects of nuclear fuel processing and handling in the Soviet Union.

For the twenty-first century, a related problem will be the disposal of wastes resulting from the decommissioning of nuclear reactors. Commercial reactors have a useful life of perhaps thirty years, after which they must be shut down ("decommissioned"), and either the entire site must be permanently isolated from human contact, or the plant must be dismantled with the radioactive remains safely isolated. Either option is expensive and not without risks, but will become a common problem in all nuclear nations as the early (1960–70) reactors become obsolete during the 1990s. The impending shut-down of some older Soviet military weapons reactors was announced in 1989.

Prior to 1986, the Soviet Union had committed itself to constructing new atomic power plants close to several major cities for the purpose of providing district-wide space heating. Such plants were originally planned for Gorkiy, Odessa, Kharkov, Minsk, Voronezh, Arkhangelsk, and perhaps several other cities as well. The Gorkiy plant is farthest along in construction, with the first of two 500 MW units scheduled to go on line in 1989; when completed it will heat the dwellings of up to 350,000 persons (Mitenkov *et al.*, 1985). The Voronezh plant is also well along in construction. However, the Chernobyl explosion may make it hard for the State Committee for

Atomic Power Safety to persuade the residents of those cities that major nuclear accidents are not possible there.

Indeed, considerable rethinking of the Soviet nuclear program appears to be occurring. In recent years, several reactor units, and even entire power station sites, have been cancelled; this had never happened prior to Chernobyl. Immediately after the Chernobyl accident, units 5 and 6 at that site, on which construction had already started, were abandoned. In 1990, the Ukrainian government directed that *all* units at Chernobyl be shut down permanently by 1995. In 1987, all units of the new Krasnodar atomic power station were cancelled, presumedly due to seismic safety reconsiderations; much site work had already been carried out there as well. Following the tragic Armenian earthquake in 1988, it was decided to commence permanently closing the Metsamor facility, even though it had not been damaged. Other proposed reactor sites, such as Kremenchug, Azerbaidzhan, and Georgia, as well as the planned Kharkov, Minsk, and Odessa heat-and-power stations were likewise cancelled (*Soviet Geography*, 30, p. 340). Some of these stations will be constructed as gas-fired conventional power stations. Other announced cancellations include the third unit at Ignalina, three units at the South Ukraine site, and apparently all units at the Crimean station.

In almost all of the post-Chernobyl closures, there had been well-reported public opposition to the proposed plants. In part, this may be due to *glasnost'*-inspired debates in the Soviet press; a small fire in unit 2 at Lithuania's Ignalina plant in 1988, for example, was well reported, and may have led to the cancellation of the third unit there. On the other hand, a degree of pro-nuclear backlash may also be developing, with nuclear proponents writing to newspapers complaining that these plants were forced to be closed by irrational and uninformed fears (e.g., *Koms. pravda*, Jan. 27, 1988).

A case could also be made that the recent cancellations were made more for economic and political reasons than for environmental ones. Some of the proposed units were in regions of questionable need, and capital resources are always scarce. By cancelling units, money can be saved (or at least redirected), public unrest can be abated, and time can be gained to replan the country's nuclear future. Later, when memories of Chernobyl are dimmer, if more nuclear power is deemed essential (and it may be), new sites can be selected. The alternative scenario is that anti-nuclear feeling is so high in the USSR that few if any new sites for atomic power stations could ever be agreed upon.

Prospects for the future

Chernobyl notwithstanding, Soviet planners envision additional increases in nuclear generating capacity during the 1990s. From a level of 12,700 MW of commercial nuclear power installed in 1980, capacity is expected to expand beyond 40,000 MW by 1990. This, however, is much less than was originally planned. In 1981 it was envisioned that 10,000 MW would be added annually during the 1986–90 five-year plan (*Pravda*, June 4). Reality will be well under half of that. Although year 2000 target figures have not yet been set, a similar cautious pace of starting up new reactors is expected (Table 3.1).

Further development of non-electrical uses of atomic energy is also likely. These include nuclear powered icebreakers and commercial ships, which the Soviet Union has operated for years (but see the discussion of the strange odyssey of the *Sevmorput* in chapter 14). Underground explosions have been used to stimulate oil and gas production, and to create storage caverns for natural gas near the city of Astrakhan. They have also been used extensively for the seismic exploration of deep geological formations (Scheimer and Borg, 1986). Between 1965 and 1983, the USSR detonated over seventy nuclear devices for non-weapons purposes (Browne, 1983). It is also possible to use such explosions to create sub-surface cavities for storing radioactive wastes, though it is not known if the USSR has ever done this. In another non-electrical application, waste heat from commercial nuclear power stations is being put to use at Kursk, South Ukraine, and other sites to raise fish and hothouse vegetables.

In summary, the Chernobyl accident has not deterred the Soviet Union from its longstanding commitment to the rapid expansion of its nuclear industry. A sizable number of Soviet officials apparently remain convinced that the safety of their commercial nuclear program can be guaranteed, though instilling this conviction in the Soviet public may not be so easy. Their standardized VVER reactors are being mass-produced at the Atommash factory, and it has been stated that they will be built with containment domes. The Soviet Union will also continue to rely on both breeder facilities and graphite moderated reactors.

With regard to long-term environmental planning, a fundamental question relates to global warming. Can fission reactors replace fossil fuel plants, and thereby decrease the release of carbon dioxide into the atmosphere? This was originally a hope within the USSR. However, Chernobyl, problems at Atommash, public opposition, and general industrial stagnation has at least greatly postponed, if not extinguished, this hope. The cancelled nuclear facilities in Azerbaidzhan, Minsk, and Krasnodar will be replaced by fossil fuel power plants. Not before some point in the twenty-first century, if then, will nuclear power stations decrease the number of carbon fuel plants in the USSR, and the amount of atmosphere transforming emissions they produce.

In the long run, as fission nuclear plants are decommissioned, it would seem desirable to replace them with fusion plants. Although commercial fusion reactors are still decades away, their advantages (abundant fuel, lack of high-level radiation, and inherent safety) justify heavy budgetary outlays for continued research. Fusion energy appears to be the only option that can produce vast amounts of electrical energy twenty-four hours a day in all parts of the country, with neither carbon dioxide nor high-level radioactive wastes as by-products.

Although the Soviet nuclear program differs from the American, French, and British in numerous respects, the radioactive residues inherent in the industry present the same potential problems in all countries. Among the many lessons of Chernobyl, certainly one of the most prominent is that the issue of nuclear safety is an international one. The imperatives of the safe operation of atomic power plants, and the safe handling of their toxic wastes, will require all nuclear nations to devote increased amounts of both planning and monetary resources towards achieving these ends.

4 Renewable energy resources

At some point in the future, the world's economy will run largely on non-fossil energy resources. This can be safely predicted on the grounds that, in the long run, there will be little choice. Not only are all fossil fuels finite in quantity, but given the problems of CO_2 build-up, global warming, acid rain, and other atmospheric problems, it is arguable that all national economies will eventually need to convert substantially to non-polluting forms of energy production.

The potential of renewable energy is great. Some individual types of renewable energy, such as solar, tidal, or the energy in ocean waves could by themselves, were their energy easily harnessable, supply all the world's energy demands (Pryde, 1983a). The problem is that few of these are economically exploitable on a large scale at the present time. But some do have promise presently, and the Soviet Union, like other developed countries, recognizes the need to investigate and develop these potential energy resources.

The types of renewable energy that are suitable to any given country depend upon that country's relative geographic location, especially its latitude, and upon its topography and geology. The USSR is so large that the many natural regions within it have differing inherent advantages *vis-à-vis* the various types of renewable energy. Each type of alternative energy will be looked at in turn concerning its feasibility in the Soviet context. By far, the type of renewable energy that has been most widely employed in the Soviet Union to date is hydro-electricity.

Hydroelectricity: the proven renewable

Around the world, hydroelectricity is the form of renewable energy that has been most extensively developed. Some countries, such as Norway, are almost entirely reliant upon it. The reason for this is that hydroelectric energy has a number of very attractive advantages.

Table 4.1   *Growth of hydroelectricity in the USSR*

Year	Hydro-electric; m.KWh[a]	Total electric; m.KWh[a]	Percent hydro-electric
1940	5,200	48,600	10.7
1950	12,700	91,200	13.9
1960	50,900	292,000	17.4
1970	124,000	741,000	16.7
1980	184,000	1,294,000	14.2
1989	224,000	1,722,000	13.0

[a] m.KWh = million kilowatt-hours
Sources: *Narodnoye khozyaystvo SSSR v 1985 g.*, 1986, p. 155; and *Soviet Geography*, Apr. 1988, p. 448

The primary advantage of hydroelectricity ("hydro") is that its input resource, rainwater is infinitely renewable (although in some areas it can be periodically unreliable). On large rivers, with which the Soviet Union is well endowed, it can produce very large quantities of power. Being based on precipitation, it is also free, and although the cost of transforming it into electricity (dam and powerhouse) are high, operational costs are very low. As a result, hydro is generally the cheapest of all technologies for generating electrical energy.

Hydroelectricity is free from all forms of air pollution. It does involve certain types of environmental problems, however, which will be discussed below in their Soviet context.

Hydro also has two technical advantages. First, it has a high conversion efficiency; that is, almost all the potential energy in the water stored behind the dam can be converted into electricity. Second, it is a very good way to provide peaking power. This refers to the extra electricity needed at times of peak demand during the day, and in terms of marketing is the most valuable type of electricity.

For all of these reasons, the development of hydroelectricity has been very popular in the USSR, and indeed that country is second in the world only to the United States in hydroelectric production. In 1989, this production equalled 224 billion kilowatt-hours, or 13.0 percent of the country's total electrical output. In 1960, hydro's share of the total had been as high as 17.4 percent (table 4.1).

It would certainly be valid to argue that the large-scale development of hydroelectricity has spared the USSR considerable additional air pollution from fossil fuel power plants, and additional nuclear plants.

On the other hand, hydroelectric dams and reservoirs also pose certain disadvantages to the environment. The creation of large reservoirs entails the loss of biotic resources, including trees, fertile soils, and valuable riparian wildlife habitat. Level land adjacent to the reservoir may become waterlogged, and fish spawning grounds may be lost. The river flow regimen will be significantly altered: less water will flow annually in the river below the dam (although its seasonal regime may be improved), sedimentation and erosion rates may be adversely affected, and salt water may encroach into coastal aquifers. Large reservoirs may produce local downwind microclimatic changes. Even small earthquakes can be induced. Most of these problems have occurred in the Soviet Union, and the result has been what is known as "the reservoir problem" in the USSR.

These problems are of particular concern in the Soviet Union, because of the large size of many of their reservoirs. Those on the Volga, Dnepr, Angara, Yenisey and elsewhere are often referred to as "inland seas," with good justification (Figure 4.1). These artificial lakes on the Volga River are so large that storms can produce ocean-like waves upon them, which easily erode the fine, wind-blown soils that line the shores. When this process undercuts trees, or creates shoals, navigational hazards result.

In a country that has difficulty meeting its agricultural goals, reservoirs have inundated millions of hectares of potentially arable land. This problem recently reached serious proportions in the Kuban, near the major city of Krasnodar. To benefit navigation and rice cultivation, the Kuban reservoir has sometimes been filled above its normal maximum level. The result was the ruin of over 100,000 hectares of crops, and water damage to 130 communities, 27,000 homes, 150 kilometers of roads, and even to the Krasnodar airport (Dergachev, 1989).

Where a purpose of the dams has been to redirect water for irrigation, downstream effects on both water quality and fish resources have sometimes been severe. For example, diversions from the Don River have reduced fresh water flows into the Sea of Azov, allowing this shallow water body to become more saline which in turn has gravely reduced the take from this extremely important fishery (Tolmazin, 1979). Although fish stocked in the reservoirs may make up for the riverine species lost to the dams, the species that have been adversely affected, such as sturgeon, are often the more valuable (chapter 10).

In response to these various reservoir problems, the USSR has for

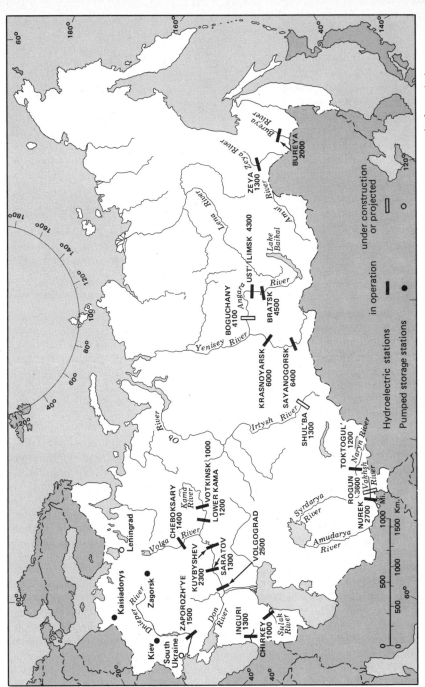

Figure 4.1   Major hydroelectric dams in the Soviet Union (numbers refer to installed megawatts at each dam site)

Figure 4.2   Wave action barrier built along shore of Kuybyshev Reservoir

years conducted continuing studies on mitigating the adverse effects of reservoirs, and has spent many millions of rubles carrying them out. Of particular concern are shore erosion problems caused by storm-generated wave actions (Figure 4.2). This problem has been especially acute on the large Volga reservoirs, as well as on some of the larger Siberian ones such as Krasnoyarsk (*Izvestiya*, Jan. 21, 1975). By the 1980s, some open questioning of the wisdom of creating such very large reservoirs began to appear. In addition to environmental problems, the extremely high cost of these projects has been a subject of concern for years.

A turning point may have been articles that appeared in *Izvestiya* and *Literaturnaya gazeta* in October of 1984, specifically pressing this question (Podgorodnikov, 1984; Zalygin, 1984). The *Izvestiya* article lamented that:

It's difficult to name a reservoir where all parameters correspond to the design, and all the consequences of the reservoir's construction have been provided for. Who foresaw ... at the Kakhovka Hydroelectric Station's reservoir the overgrowth, the water bloom, and all the undesirable conditions that were created there? ... Who anticipated the sharp falloff in the rated output of the Novosibirsk Hydroelectric station on the Ob', the losses to agriculture on the Don or the losses to fisheries on the Volga? (Zalygin, 1984)

In subsequent years, more questioning articles appeared (Paton, 1986; Avakyan, 1987). Among many unfortunate examples, Academician Paton cites the Kiev reservoir, which at around 1,000 sq. km can be viewed as excessively large. A plan to dike off part of it was rejected for economic reasons. As a result, vast areas of fertile floodplains were inundated. Additionally, the shallow reservoir heated up in summer, producing extensive algae blooms which deoxygenated the water, killed off fish, and caused recreational values to deteriorate.

With the advant of the *glasnost'* era in the mid-1980s, specific public opposition to particular proposed dams arose. The giant Rogun Dam, now under construction in Uzbekistan, for example, has recently been questioned regarding its seismic safety (*Pravda*, Nov. 21, 1988). The project that has perhaps drawn the most criticism is the proposed Katun hydroelectric station. Located on the upper Ob' River in the Altai mountains southeast of Barnaul, the main dam would generate 1,570 megawatts of electricity. But, around 1984, shortly after site preparation began, the project's effects began to be questioned (Sapov, 1984; Cherkasova, 1990). At a regional conference in 1986 on the environmental effects of the dam (which the project designers declined to attend), serious questions were raised concerning flooding problems, site hydrogeology, microclimatic effects, and earthquake safety (Vinokurov, 1986). A series of articles on the merits and problems of Katun also appeared in the journal *Priroda i chelovek* (Nature and Man) in 1989 (Figure 4.3). Despite this, the project continues under construction.

As the Soviet Union enters the 1990s, both environmental and economic factors appear to be resulting in a reconsideration of the merits of large conventional hydroelectric projects. However, there are also some attractive, non-conventional ways of producing electricity from moving water that can be less environmentally damaging. These include pumped storage plants, tidal plants, and small hydroelectric facilities.

Pumped storage plants are a very attractive means for providing power for peak demand periods. They utilize inexpensive night-time electricity from fossil fuel or nuclear power plants to pump water uphill to an artificial reservoir, and then run it downhill to create more valuable peak period electricity the next day. The lower water body can be any large river or lake, which does not have to be dammed; only an elevation difference and an existing electricity source are needed. A great many nations are now building this type of plant.

In the Soviet Union, pumped storage plants have been built near

Figure 4.3   Cover art opposing the Katun dam *Source:* journal
Priorda i chelovek (Nature and man)

Zagorsk, at the South Ukraine atomic power station, near Kiev, and at
Kaisiadorys in Lithuania. The Zagorsk plant, the USSR's first, has a
capacity of 1,200 MW, the South Ukraine 1,800 MW, and the Lithuanian
one 1,600 MW (*Pravda*, Sep. 29, 1982). More are under construction, or
planned, in Armenia, the Tatar autonomous republic, the Ukraine
(two sites), near Leningrad, and elsewhere (Kelly *et al.*, 1986, p. 217;

Soviet Geography, 1986, p. 3). These plants are logical and environmentally benign compliments to nuclear power stations, and the 1990s will probably see many more of them built in the European part of the USSR.

The attraction of tidal power is its vast potential output; it is one of those forms that, if harnessable, could supply all of the world's energy needs. It has significant disadvantages, though, including coastal disfiguration, problems of working with saltwater, and generally higher costs per kilowatt than conventional hydro plants. The Soviet Union has shown some interest in tidal plants, and placed a small, prefabricated unit (1,000 KW) into service at Kislaya Bay on the Kola Peninsula as early as 1968 (Pryde, 1972, p. 131). Although much larger plants have been discussed for other nearby sites, such as Lumbovskaya Bay and Mezen Bay, no construction had started as of 1990. This option now appears to have a low priority.

A final way to produce small amounts of hydroelectricity is to install small turbines in *any* source of moving water. This can include irrigation canals, pipelines conveying municipal water supplies, or even wastewater lines – any channel or pipeline that has an elevation drop. This option seems to be little used in the USSR, probably because of the preference of Soviet planners for large projects, but occasional calls for this kind of retrofitting are heard (*Pravda*, June 12, 1983). Experience in the United States and elsewhere has shown that these small hydro applications on existing facilities can often be highly cost effective.

Soviet development of solar energy

Despite its general worldwide popularity, solar energy research in the Soviet Union through the 1970s received only modest emphasis (Grunbaum, 1978; Pryde, 1978; Campbell, 1980). This reflected an economy that had been based on abundant fossil fuel resources, augmented by vigorous nuclear and hydroelectric energy development (Dienes and Shabad, 1979; Campbell, 1980; Kelly *et al.*, 1986). A 1981 Soviet book on energy devotes only 6 of 304 pages to a brief outline of solar technologies, and wind received just 1 page (Venikov and Putyatin, 1981). Even in a book specifically on alternative energy a similar slighting of solar (7 pages out of 62) is found (Marochek and Solov'yev, 1981). The amount of Soviet solar research in 1975 was estimated at only about 5 million rubles (Campbell, 1980, p. 52). Although a much greater amount was being spent in the mid-1980s,

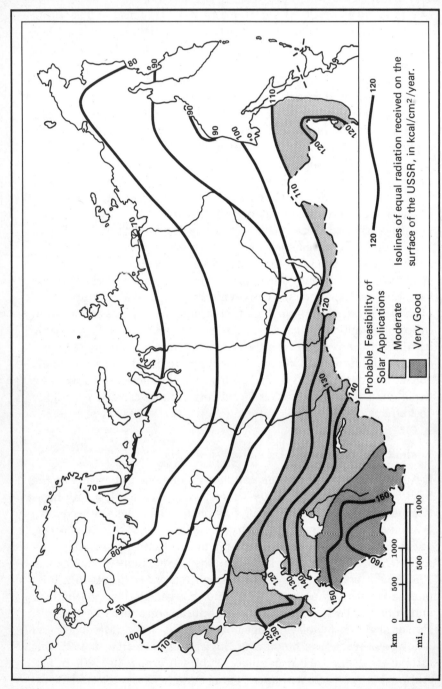

Figure 4.4 Mean annual total solar radiation for the USSR *Source*: Lydolph, 1977, as adopted from Pivovarova, 1966

Table 4.2   *Areal potential for solar energy*

Radiation received	Approximate percent of country	
	US (48 states)	USSR
110 kCal/cm²/yr[a] (300 L./day)	96	25
146 kCal/cm²/yr[b] (400 L./day)	46	6

Notes: [a] Approximate minimum level for commercial feasibility of solar technologies
[b] More efficacious level for commercial feasibility of solar technologies

there was little commercially available solar equipment to show for it (Konovalov, 1983). Soviet specialists themselves seem to confirm that the USSR is not a leader in solar research: in a recent article on the country's solar energy potential, only 5 out of 29 citations were to Soviet authors (Reznikov, 1989).

The low Soviet interest in solar energy in part reflects the funda-mental reality of the northerly location of the USSR. In general, only the southernmost portions of the USSR are well suited to solar energy utilization, especially in winter when supplemental heat and elec-tricity are most needed (Figure 4.4). Summer months benefit from long days and high sun angle, but it is also the season of maximum precipitation in most parts of the USSR. In the United States, average solar radiation of 400 Langleys/day is considered desirable for effective direct solar applications, and about 46 percent of the conterminous states receive this much. Four hundred Langleys/day equals 146 kCal/cm²/yr, a figure achieved in the Soviet Union only in Central Asia below a latitude of about 43 degrees, or on about 6 percent of the USSR's territory (Table 4.2).

Solar energy also has some inherent drawbacks. Although it is an almost infinitely renewable resource, it is intrinsically limited by being both diffuse and intermittent, and therefore must generally be conver-ted into forms that are more reliable and concentrated. Nevertheless, faced with fossil fuel resources that are neither inexhaustible nor inexpensive, the Soviet Union has stepped up its investigations into the solar alternative.

Solar research in the Soviet Union began in the 1930s, and is today carried out at a large number of laboratories and field stations. The most elaborate of these facilities is the Solar Energy Research Institute at the new town of Bikrova, near Ashkhabad (Presnyakov, 1980). Soviet research into solar energy is reported in the periodical *Geliotekhnika*, which is available in translation as *Applied Solar Energy*.

Among small-scale solar applications, one of the most emphasized has been the distillation of saline water. Numerous solar stills have been constructed in arid regions, the largest of which, at the Bakharden Sovkhoz in the Turkmen republic, can produce 1.9 million liters (500,000 gallons) of fresh water per year (Pryde, 1984a). Solar cells are used to pump the water.

Systems to heat water are one of the most common solar applications, and have been put into operation at several locations in the Crimea, Central Asia, and elsewhere. Solar water heaters and ovens are produced at a solar equipment factory in Bukhara, but apparently in relatively small quantities.

Although roof-top collectors for providing residential hot water are commonplace in many parts of the United States, Australia, and elsewhere, this is evidently not an emphasis in the USSR. Development through the mid-1980s had been limited to demonstration projects. In their book on energy technology, Venikov and Putyatin dispose of both domestic hot water and space heating in four sentences, and no indications of official encouragement for such applications have been noted. The same appears to hold true for passive heating techniques for buildings, although some research on passive systems is carried out at Bikrova (Presnyakov, 1980).

Soviet physicists seem to have a strong interest in solar refractory furnaces, which use intense solar rays to liquefy metals and other substances having a high melting point. Experimental units were first built in Armenia and the Crimea, and a larger (1 MWth) solar furnace that produces 3,000 C heat is being constructed near Tashkent (Azimov, 1987).

Sun-tracking mirror systems that heat water or other fluids can be used to produce electricity, and these appear to be the favored design for solar electrical plants in the USSR (Aparisi *et al.*, 1980). The largest Soviet facility of this type, located near the Crimean city of Kerch, is a 5 MW heliostat array whose 1,600 mirrors can generate up to 12.5 million KWh of electricity a year. The first phase of this plant was completed in 1986. Studies are being conducted on a much larger system that could generate 300 MW, to be built in the Uzbek republic (*Izvestiya*, Feb. 10, 1985, p. 1).

Photovoltaics refers to the direct conversion of solar energy into electrical current. Although photovoltaic cells (or solar cells) have been used in the Soviet space program since Sputnik 3, the Soviet Union has not been a leader in this type of research. The use of solar cells for navigation, water pumping, and beacons, is common in remote desert and maritime areas. A 100 KW power plant is being built at the Bikrova research center, with bigger ones proposed (Presnyakov, 1980), but much larger facilities are currently operational in other countries. The potential of photovoltaics can be seen from the estimate that all of the USSR's 1980 electrical demands could be met by solar cells covering an area in Central Asia just 65 km per side (Alferov, 1983).

Orbiting solar cell satellites of huge size that would beam microwave energy back to earth have been studied in both the United States and USSR. In the past, Soviet planners tended to favor large-scale projects of this type, and it is noteworthy that almost half of the short solar section in the Venikov and Putyatin book is devoted to energy beams from satellites; a similar prominence is found in the more recent work by Nikitin and Novikov cited in the bibliography. These energy satellites are generally dismissed in the United States as prohibitively expensive.

For the future, despite talk of constructing large centralized solar power stations of up to 300 MW, it appears that the bulk of Soviet efforts will remain in the practical development and application of small, simple, and proven solar pumping and heating systems. The construction of large centralized solar power stations seems unlikely under tight *perestroika* budgets, although the idea might be defensible in Central Asia. For the 1990s, continued research, plus the production of various small-to-medium sized solar devices and demonstration projects, appears to represent the probable extent of Soviet interest in the solar energy alternative.

Other types of renewable energy

Wind Power. Substituting the pollution-free energy of the wind for power derived from fossil fuels enjoyed great popularity in the 1980s in many parts of Europe and North America. Wind energy resembles solar in being infinitely renewable, but it is also similarly diffuse and intermittent, and its large-scale development for electrical generation requires many massive structures.

On the other hand, wind power has been highly feasible for centuries as a small-scale source of mechanical energy. In the Soviet Union, as elsewhere, there are thousands of windmills of this type

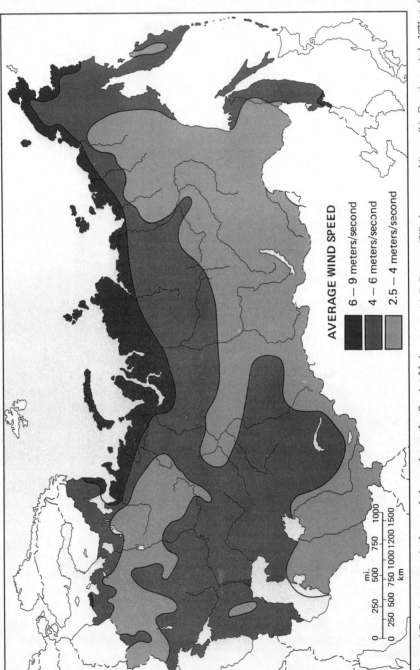

AVERAGE WIND SPEED

6 – 9 meters/second

4 – 6 meters/second

2.5 – 4 meters/second

mi.
0   250   500   750 1000 1200 1500
km
0   250   500   750 1000 1500

Figure 4.5   Average wind speed throughout the Soviet Union *Source:* M. E. Fateyev, "Wind energy and its use," in Gerasimov et al., 1971, p. 95

used for such tasks as pumping ground water. In rural areas, they may always be useful for such functions.

The production of electrical energy from the wind has apparently received less attention. This is in a sense ironic, since the USSR has attractive areas for wind development (Lydolph, 1977; Figure 4.5), and also because prior to World War II the USSR was a leader in this field. A wind-powered electrical generator of 100 KW was built in the Crimea in 1931, and by 1938 plans were underway for a much larger one of 5,000 KW. Both projects were terminated by the war. During the past few decades, wind machines with a variety of generating capacities up to 100 KW have been built for use on collective farms and in other rural locales.

In general, however, progress has been slow. Research and development on wind generators is carried out by the "Vetroen" (earlier called "Tsiklon") association of the Ministry of Reclamation and Water Resources. Although this organization was created in 1975, *Izvestiya* noted that by 1983 its Astrakhan assembly plant had produced only demonstration models, not any integrated production facilities, despite the expenditure of 23 million rubles (Konovalov, 1983). Other articles, similarly critical, have appeared. Starting in 1984, Vetroen was supposed to be producing 200 operational units per year. These smaller, agriculturally oriented units appear to be the main focus of Soviet wind energy development; larger units of megawatt-size output seem to have a very low developmental priority.

Geothermal energy. Geothermal ("earth heat") resources represent an attractive non-conventional energy source that over the past three decades has received a great deal of worldwide attention.

The Soviet Union has many areas with hot springs or other surface manifestations of geothermal energy. The geyser and volcano region on the Kamchatka Peninsula is the most spectacular, but hot springs are also common in the Trans- and North Caucasus, in the mountains of Central Asia, and in various locations throughout Siberia. Wide use is made of these heated waters for resort and health spa purposes, especially in the Caucasus.

In addition to health spas, most of the practical development of geothermal energy in the Soviet Union has focused on space heating, water heating, and agricultural applications in areas such as the North Caucasus, Georgia, Siberia, and the Far East. In the city of Makhachkala on the Caspian Sea, 60 percent of its hot water demands are met by fluids from geothermal aquifers. In the Georgian capital of Tbilisi, 25,000 persons benefit from geothermal heat, and a few smaller towns have been entirely or mostly supplied from this source since the 1970s. Geothermal fluids are used to heat greenhouses in many locations in the North Caucasus, Georgia, Kazakhstan, near Ulan-Ude, and on Kamchatka (Pryde, 1979). However, there have been complaints that such development is moving very slowly (*Pravda*, Dec. 7, 1983, p. 3).

Given the Soviet Union's northerly latitude and widespread low-temperature geothermal resources, it is probable that its geothermal research activities will continue to focus on low temperature heating and agricultural uses. These will probably remain the center of attention at least through the 1990s.

Geothermal steam has been used since 1904 to generate electricity, and about a dozen countries, including the USSR, have utilized this process. In 1967, a 5 MW geothermal generating station was completed at Pauzhetka near the southern tip of the Kamchatka Peninsula. Over twenty wells have been drilled at this site, and its capacity was reportedly doubled in 1981 to 11 MW. A smaller plant, using a freon heat-exchange system, operated from 1968 to 1975 at Paratunka, near the city of Petropavlovsk. Many other proposals for developing Kamchatka's resources have been put forth (Dvorov, 1986). However, twenty years later, no large geothermal electrical facilities have yet come on line in the USSR, although there has been talk of building one in Dagestan (*Izvestiya*, Aug. 24, 1982, p. 3). At present, the Soviet Union cannot be termed a world leader in geothermal powerplant development.

Other non-conventional energy sources. Several other potential alternative energy sources exist. Although some are being used elsewhere in the world, none of them are currently important in the USSR. A very brief review of them follows.

Energy can be extracted from the oceans through waves, currents, and vertical heat gradients. The USSR has no off-shore areas with warm enough waters to make the temperature gradient concept feasible, and little has been seen on using currents. However, an article in the March 1988 issue of *Alternative Sources of Energy* noted that some research was being conducted on converting ocean wave energy into mechanical rotational energy. Apparently, though, deployment of even a pilot-sized facility is still years away.

The conversion of trees, cultivated plants, or other organic matter ("biomass") into secondary fuels, such as methane or alcohol, is becoming a common practice in many countries. It can also be controversial, because of the environmental or food production implications of converting existing forests or agricultural soils to these purposes. Research is conducted in the USSR on many biomass conversion processes, but none have as yet been developed on a large scale. One related technique, the processing of agricultural or forestry wastes to produce natural gas or electricity, is less controversial and is also being done on a small scale.

An interesting secondary fuel is hydrogen, which can be burned in such a way as not to produce carbon or hydrogen based pollutants (its by-product is water vapor). Its use as a fuel would still produce nitrogen oxides, however. In April of 1988, the Soviet Union reportedly flew an airliner powered by hydrogen for the first time. It is clear that hydrogen is a desirable fuel, and that the USSR is interested in developing it; what is not now clear is the economic feasibility of producing hydrogen as a fuel in huge volumes (Kelly et al., 1986, p. 285).

Certain other types of nonconventional fuels or conversion techniques, such as oil shale, coal liquefaction, tar sands, and the above mentioned alcohol fuels, are either not used in the USSR, are so controversial that their use is unlikely to be expanded (e.g., oil shale), or are debatable as to whether their large-scale development would result in any net environmental gains, when analyzed in a comprehensive manner (Pryde, 1983). The USSR has shown some interest in coal gasification, however, and a great deal of interest in MHD (magnetohydrodynamics) techniques, with an MHD plant now operational at Ryazan.

A renewable future for the USSR?

What role will non-conventional energy play in the Soviet Union in the next century? Is there a renewable energy future for the USSR?

If these questions are asking whether such sources will account for a majority of Soviet energy production in the foreseeable future, the answer is almost certainly "no." However, renewable energy will play a role, and in the long run perhaps even a significant role.

Hydroelectricity will continue to dominate the renewables, though the case for more hydroelectric power is debatable (Avakyan, 1987). Hydroelectric dams destroy river valleys, mineral and biological resources, and agricultural land, but they embody all the advantages enumerated earlier. As atmospheric changes become of more concern, the arguments for relatively clean hydroelectricity may seem more compelling. In particular, pumped storage plants are an attractive option for the USSR, especially if they continue to expand their nuclear program.

Hydroelectric energy should probably *not* be expanded in Central Asia and the Transcaucasus, as in these drier regions the riparian resources of river valleys are extremely important. Fortunately, solar energy is feasible in these more arid regions. The development of solar energy in Transcaucasia and Central Asia (including Kazakhstan) should be aggressively pursued.

Most of the other forms of renewable energy are unlikely to see large-scale development. Low average wind speed will limit electrical production from that source, and ocean thermal gradients of sufficient temperatures do not exist in Soviet waters. Both wave and tidal power stations are possible, but as of 1990 must be considered unlikely choices. Biomass conversion into either alcohol fuels or electricity, except possibly from waste feedstocks, is not probable on a significant scale. Geothermal energy will remain locally important for space heating and greenhouse applications, but will account for very limited electrical production. The future of hydrogen as a secondary fuel is harder to predict, but does not seem at present to be a priority development in the USSR.

Holding back the pursuit of all these options is a meager level of funding, apparently reflecting a low official priority. Konovalov, writing in 1983, noted that the first All-Union Conference on Renewable Energy Sources had been held in Tashkent in 1972, but that there had never been a second one. Interest may be picking up, however, as the slow pace of renewables was discussed in the CPSU Central Committee in 1988 (*Pravda*, June 9, 1988; Appendix 4.1). In the early 1980s, a coordinating body called the All-Union Scientific Council on Problems of Using Renewable Energy Resources was created.

In summary, renewable sources of energy are most unlikely to meet all, or even the majority, of future Soviet electrical needs, but they can be, and should be, locally important. The solar energy potential of Central Asia, in particular, would appear feasible and desirable to emphasize in future five-year plans. An increased emphasis on appropriate renewables, coupled with nuclear energy and above all a much higher priority on the more efficient use and conservation of presently available energy, should permit an eventual reduction in carbon dioxide and other harmful fossil fuel plant emissions into the Soviet Union's (and world's) already troubled atmosphere.

Appendix 4.1
Excerpt from: "Basic provisions of the USSR's long-term energy program"

NONTRADITIONAL RENEWABLE ENERGY SOURCES AND SYNTHETIC LIQUID MOTOR FUELS

In the first stage of the implementation of the USSR Energy Program, it is planned to create a material and technical base for the broad use of nontraditional energy sources – solar, geothermal, wind and tidal energy and biomass – as well as to solve basic scientific and technical problems in the field of the production of synthetic liquid motor fuels from gas, coal and oil shale.

In the second stage, plans call for beginning the active drawing into the energy balance of nontraditional production of synthetic liquid motor fuels.

The most likely field for the use of solar energy in the national economy will be low-temperature heat supply in the southern regions of the country.

By the end of the second stage, the annual production of energy resources through nontraditional energy sources will total between 20 million and 40 million tons of standard fuel. The bulk of these resources will be obtained from the use of solar and geothermal energy, as well as from biomass.

Research and experimental-design work in the field of solar power engineering and the use of the earth's heat will be linked to improving the efficiency of photoelectric transformers of solar energy, perfecting the configurations and designs of solar units operating on the heat cycle and optimizing solutions in the field of the use of geothermal water for the production of heat and electric power, as well as to preparing the scientific and technical base for the development of petrogeothermal energy resources. A future increase in the economic indices of geothermal power engineering will be ensured in large part by the development and introduction in production of effective technologies for the comprehensive utilization of geothermal water, with the recovery of the useful components contained therein.

In the second stage, the construction of the first industrial enterprises for the production of synthetic liquid fuels from coal will begin in the Kansk-Achinsk Coal Basin. A great deal of attention will be devoted to the development and introduction of new methods of liquefying coal that will make it possible to substantially increase the unit capacity of technological installations and the direct processing of methanol into motor fuel, as well as the use of hydrogen as a motor fuel.

Source: Ekon. Gaz., March 1984, no. 12, pp. 11–14, as translated in Current Digest of the Soviet Press, 36 (1984), no. 24, pp. 12–16

5   The quest for clean water

Water pollution, like deterioration of the atmosphere, is an inevitable offspring of industrialization. With the growth of industry in the Russian Empire prior to the Revolution, and far more so in the Soviet period that followed, it could be predicted that numerous rivers and lakes would pay the price. For decades Marxist ideologists argued naively that the planned nature of socialist societies would largely "preclude" significant pollution, and academicians not only echoed them but compounded the problem by alluding to water as an "inexhaustible" resource (Pryde, 1972; ZumBrunnen, 1984). Some Soviet scientists even seemed to deliberately overestimate the self-purification role of rivers and lakes, so as to justify fairly high "permissible" levels of pollution (Pryde, 1983). Fortunately, a more realistic assessment of the causes, and cures, of pollution is gaining acceptance in the Soviet Union today. More data is also available. Indeed, the 1989 statistical handbook shows a startling increase in polluted water discharges in the late 1980s: from 15.9 billion cubic meters in 1985 to 20.6 in 1987, to 28.6 in 1988 (*Okhrana*, 1989, pp. 78–80). An increase of 80 percent in four years begs for some elaboration but unfortunately no explanation for the rapid increase is given.

Common sources of water pollution in any society are cities, agriculture, and such inherently "dirty" industries as mining, steel, petrochemicals, pulp-and-paper, nonferrous metals, and textiles. To counter such threats, the Soviet government has devoted much energy and money to correcting their internal water problems, a reported 8.1 billion rubles during the 11th five-year plan (1981–85) alone. However, this figure includes some irrigation and reclamation projects, as well as wastewater facilities. ZumBrunnen, in his 1984 review article, cites numerous Soviet reports from the 1976–82 period that describe both critical pollution problems and concerted ameliorative efforts (pp. 266–70), as did earlier works for the 1960–71 period

Figure 5.1 Anti-water pollution magazine cover. *Source: Priroda i chelovek* (Nature and man)

(Pryde, 1972; Goldman, 1972). Industrial expansion, it seems, frequently means that as fast as purification progress is made in one river or lake, equally serious problems are identified at another.

There is also reason in certain areas to be concerned about the quality of municipal drinking water (Figure 5.1). The water supply of Kharkov (the Northern Donets River) was prohibitively contaminated in the 1970s, the water supply for Kiev was at best in severe peril following the Chernobyl episode, and the drinking water in Leningrad is known to have been fouled by pollutants from upstream sources for years. Maximum Permissible Concentrations have been established for 420 pollutants in potable water supplies (Jancar, 1987, p. 112), but the above gives reason to question the extent of their compliance.

A particularly serious health problem appears to exist in the Turkmen republic with regard to public water supplies. A lengthy article in the newspaper *Komsomolskaya pravda* on April 25, 1990, described the situation in one village as follows:

The situation today can be assessed in one word – catastrophic. Just the use for drinking of water poisoned by toxic chemicals – and in many areas there is no other kind of water! – claims hundreds of lives every year, first of all children's lives. We looked at the results of a laboratory analysis of water from wells in the settlement of Takhta, Tashauz Province. It is horrifying to think that people drink this water! The sulfates content is 50 times higher than the norm, that of chlorides 40 times higher, of calcium 17 times higher and of magnesium 10 times higher! Right now, it is not recommended that fish be caught for use as food in many canals, lakes and reservoirs. Practically all of the vegetables and fruit grown in the republic's cotton-growing zone (a large part of Turkmenia's territory) have an elevated nitrates content. A sample examination of nursing women conducted last year showed that their mother's milk contains pesticide residues! (Voshchanov and Bushev, 1990)

Rivers and residuals

Municipal wastes have long been a serious source of pollution into Soviet rivers and streams. Prior to World War II, most Soviet cities had no municipal wastewater treatment plants. Even in 1960, only 51 cities had sewage treatment facilities, and 60 percent of all municipal wastes were being discharged into natural water bodies in an untreated form (Pryde, 1972). Even in 1983, there were still 350 Soviet towns and cities lacking sewerage systems (Sebastian, 1988). As recently as 1986, only 58 percent of Leningrad's sewage was receiving treatment, and full treatment is not likely until after the year 2000 (Precoda, 1988). From the standpoint of public health, as well as proper water resources management and fisheries conservation, this was a situation that had to be corrected.

Over the past three decades, a significant emphasis has been placed on the construction of municipal wastewater treatment facilities in Leningrad and other cities. Between 1976 and 1982 capacity was added to treat over 48 million cubic meters per day of wastewater, reducing flows of untreated sewage by 25 percent (Nuriyev, 1983). However, this implies that 75 percent of the volume of raw sewage being discharged in 1976 was still flowing six years later. Unless a huge increase in funding was made available during the next seven years, the conclusion seems warranted that a significant amount of untreated municipal sewage must still be reaching Soviet rivers and streams in 1990.

The direct consequences of such raw effluent discharges are obvious, but the indirect effects can sometimes be more complex. For example, a 1984 report noted that flows of untreated sewage had eroded a ravine in the city of Dnepropetrovsk and undermined a housing complex, requiring permanent relocation of fourteen families and temporary relocation of fifty others (*Izvestiya*, Jan. 5). In addition to Dnepropetrovsk and Leningrad, it is known that Sochi, Mariupol (Zhdanov), and several other major cities still have serious sewage treatment problems.

Municipal wastewaters should receive, as a minimum, primary treatment, which removes around half of the organic materials in the wastewater flow, and preferably secondary treatment as well, which increases the organic removal up to 85–95 percent. The author in 1978 visited a treatment plant in the city of Shakhty, which handled 80,000 cubic meters of waste per day. This particular plant was equipped with secondary treatment facilities using the activated sludge process, though the exact percentage of organic materials removed was not known by the tour leaders. The plant also treated wastes from a nearby textile plant; we were told that chemicals (dyes, etc.) that could not be eliminated by secondary treatment were removed at the factory before the waste stream went to the treatment plant. If everything worked as described, and if the facility was adequately sized, it should have been capable of preventing pollution of the water bodies receiving its treated effluent. Whether or not it was typical of other treatment plants in the USSR at the time is not known; in all probability it was better than the norm.

A great variety of pollutants originate at industrial complexes, and these are not always contained or neutralized on site. Although many newer factories have exemplary treatment facilities, it is not uncommon for the portions of rivers downstream from most Soviet cities and industrial sites to be contaminated to some extent by water pollutants. As examples, oil pollution closed the Mulyanka River (near Perm) to swimming in 1987, the Nuren River in the Ukraine caught fire following discharges of oil into it in 1989, and much of the Dnestr

River was severely contaminated by the collapse of a holding dam in 1983 (see chapter 6). A number of rivers in such industrial regions as the eastern Ukraine, Ural mountains, and West Siberia have become highly polluted (ZumBrunnen, 1984).

Water samples that do not meet adopted standards are common. Within the large Russian Republic (RSFSR), a 1989 report noted that 8.1 percent of water samples taken from municipal water mains and 14.2 percent of samples taken from industrial mains exceeded the national standard for bacteria count. Further, 19.2 percent of the municipal samples and 20.4 percent of the industrial samples were above the standards for chemical content (Pipia, 1989). Cities cited in the report where the problem was especially acute included Arkh-angelsk, Astrakhan, Kemerovo, Krasnodar, Krasnoyarsk, Murmansk, Perm, and Sverdlovsk.

The fabled Volga, longest river in Europe and a national symbol of the Russian people, has had a longstanding history of pollution problems (Pryde, 1972; Shipunov, 1987; Wolfson, 1988a). Not only have its many dams blocked fish runs, but the still waters in the reservoirs are highly vulnerable to algae blooms. Concern about the Volga has been expressed since the 1960s, and in 1984 an Institute of the Ecology of the Volga Basin was established within the USSR Academy of Sciences.

The condition of the river was surveyed by a group of Soviet scientists in 1987 and 1988, and the verdict was not good. The members of the expedition reported that

the natural aquatic systems have been destroyed ... The Volga has been transformed ... into a chaotic mess. Some 300 million tons of solid matter is dumped into the Volga annually. Water transparency has dropped from several meters 50 years ago to about one-half meter today. ...

Polluted industrial discharges combine with agricultural runoff containing toxic chemicals and residues from mineral fertilizers. The Volga's annual flow is insufficient to render harmless the thousands of tons of phenols and other biologically dangerous chemicals and organic compounds, tens of thousands of tons of nitrates, and hundreds of thousands of tons of suspended sub-stances discharged every year. These poisons reach our tables in the food that is grown in fields irrigated with this water. ("Hands in the Volga," 1988)

In a *glasnost'* response to the continuing severity of the problem, in 1989 a "Public Committee to Save the Volga" was formed (chapter 14).

Agricultural activities have also caused great deterioration in other rivers. The Amu-Darya River in Central Asia is highly polluted from irrigation return flows contaminated by agricultural chemicals, compounded by a greatly reduced flow in the river due to irrigation diversions (see chapter 12). Overland (that is, non-channelized) flows from agriculture, urban streets, and spring meltwaters are a significant source of pollutant loading that is now starting to receive increased attention (Chernyshev and Barymova, 1985). In the Russian Republic alone, 900 million cubic meters of waste from cattle breeding enterprises finds its way into rivers annually (*Soviet Rossiya*, June 10, 1987). It might be asked why these methane-producing animal wastes are not being converted into biogas, a valuable economic commodity?

In the Soviet Union today, it is standard practice to require adequate purification facilities at all new industrial plants, but it is not uncommon for these factories to begin production without the mandated treatment facilities being operational (Khrenov, 1985). A similar situation is often found at older factories, where adequate funds to abate existing pollution problems, or to handle an enlargement in capacity, often are not made available.

An illustration of this type of problem occurred at Saratov, where chemical pollution is thought responsible for a sharp increase in environmentally related diseases (Wolfson, 1988a). Here, a biochemical plant had built the required wastewater treatment facilities, but upon inspection they were deemed to be poorly constructed, and were not accepted by the health authorities. However, it was also observed that they could not be used even if rebuilt, as the local wastewater line into which they would be connected was already exceeding its capacity. An article in the government paper inquired sardonically if this implied that another five-year plan period would have to pass before the problem could be cured (*Pravda*, Jan. 28, 1983).

The basic legislation governing water quality activities in the USSR was enacted in 1960. A set of normative standards for pollutants in water bodies, called Maximum Permissible Concentrations (MPC), has existed since the 1940s. These standards are typically very strict (usually stricter than corresponding American standards), and in many cases it would be difficult if not impossible to actually comply with them. Originally, the Ministry of Public Health was made responsible for establishing and enforcing water quality standards; currently the new State Committee for Environmental Protection (Goskompriroda) is the responsible agency. Created in 1988, it is too new for its performance to have produced visible results. Over the past twenty years, numerous special decrees have been enacted by the Council of Ministers that seek to respond to particular regional problems, such as improving water quality in the Volga Basin, in small rivers, in Lake Ladoga, the Desna River, the Baltic Sea, Lake Baikal, or the Caspian Sea. Some of these regional problems will be examined in more detail below.

Lakes and *laissez-faire*

Although lakes are sometimes described as "wide spots in rivers," they differ from the rivers that feed them in at least one important respect: turnover time. Turnover time refers to the amount of time that is required for inflows of new water to replace the existing stock of water in a section of river, or in a lake. It can be seen that this period of time (and the required amount of water) will generally be much greater for a lake than for a river segment of similar length. For example, a 10 km stretch of a river, even a large one, will have its stock of water completely replaced in a few hours, whereas the turnover time for a large lake is often measured in years (400 years in the case of Lake Baikal). When pollutants are released into lakes, this distinction can be seen to be highly significant from the standpoint of restoring water quality by natural flushing action. Thus, the protection of water quality in lakes everywhere is of particular importance.

The Soviet Union has countless lakes. Most, particularly in Siberia, are small and sufficiently remote that they are still fairly pristine. A great many, however, and most notably some of the larger ones, have been used as the sites of industrial activities, and have consequently suffered some degree of pollution. Among the major freshwater lakes that have been the focus of pollution controversies are Sevan, Ladoga, and Baikal.

The protection of Lake Sevan was one of the subjects of demonstrations that took place in Armenia in 1987. Its level has dropped by six meters since the 1930s due to diversions for industrial and agricultural purposes. An expensive tunnel built in the 1970s to bring water from the Arpa River in Azerbaidzhan to the lake did not correct the deficiency (and may not have helped inter-republic relations). As the lake's level fell, its water quality deteriorated proportionately. An interceptor pipeline to capture effluents originating in towns around the lake has been authorized, but its construction is proceeding slowly (Wolfson, 1988a). It is likely that financial resources for this and other projects will be diverted to help pay the 16 billion ruble cost of rebuilding cities destroyed in the December, 1988, earthquake.

Lake Ladoga has been one of the most serious problems in recent years. This large lake lies just northeast of Leningrad, and supplies the

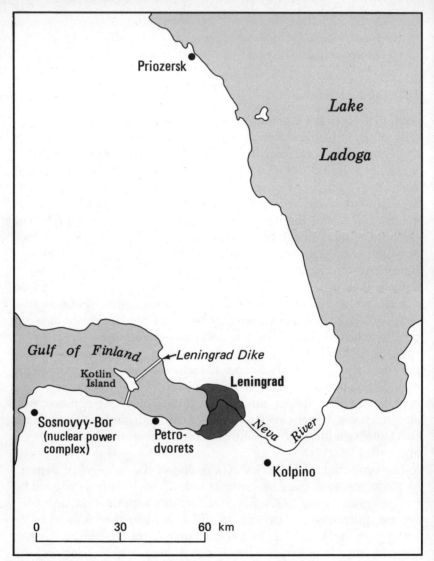

Figure 5.2   Leningrad, Lake Ladoga, and vicinity *Source: Atlas SSSR,* 1983

water both for the Neva River which flows through that city, and for the city's domestic water supply (Figure 5.2). Initial reports of serious pollution problems appeared as early as 1967, and in 1984 the USSR Council of Ministers adopted a special resolution on ensuring proper use of the water resources of Lake Ladoga and other nearby lakes.

The main focus of concern at Lake Ladoga was an old pulp mill at the town of Priozersk, which for decades had discharged untreated industrial wastes which had created a hazard to the town, to the lake's ecosystem, and to the Leningrad water supply (Precoda, 1988). After years of debate, the Priozersk pulp operations were shut down in 1986, and the factory was converted to making chipboard for the manufacture of furniture. In addition, numerous officials were reprimanded (*Pravda*, May 29, 1987). Unfortunately, other mills still dot the lakeshore, and some, such as the Svetogorsk cellulose combine, continue to discharge large volumes of polluted water into it (*Okhrana*, 1989, p. 136). In 1989 the Leningrad Environmental Protection Committee concluded that the deterioration of Lake Ladoga "had taken on a permanent character" (*Vecherniy Leningrad*, June 1).

Reservoirs are a common form of man-made lake. As noted in the preceding chapter, the multitude of reservoirs in the USSR have come under considerable scrutiny for a number of reasons, of which one of the foremost is water quality. Eutrophication, the tendency of large, slow-moving water bodies to grow algae blooms during the summer months, has been a common problem on some of the larger reservoirs such as those on the Volga River. Reservoirs also create ecological problems by waterlogging the land around them. Other problems associated with these reservoirs were discussed in chapter 4.

Thermal pollution (the heating of natural water bodies) can also cause lakes to deteriorate. As one example, part of the reason for the protests over the Ignalina nuclear power plant in Lithuania involves the discharge of the huge volumes of heated water from the reactor's cooling cycle into nearby lakes. Irreversible ecological damage to the lakes was feared by nearby residents.

The Volga flows into the Caspian Sea, which itself has suffered from a variety of pollution problems. These have originated from inflows from the Volga and other rivers, from cities and factories around its shoreline, and from many decades of oil extraction. The oil production was initially carried out along its shores, but since the 1950s has increasingly utilized off-shore drilling platforms built in the sea itself. Pollution has also entered the Caspian Sea from Iran, as in 1975 when 2 million fish were killed by discharges from a Japanese–Iranian appliance factory near the south coast. The sad fate of the Aral Sea will be examined in chapter 12.

The question of water quality in Lake Baikal has been so controversial for so long that it merits a more detailed review.

The battle for Baikal: victory or vicissitude?

Since the mid-1960s, the most significant struggle against water pollution in the Soviet Union has been the fight to protect Lake Baikal. Indeed, the beginnings of the present era of environmental awareness in the Soviet Union are usually traced to the controversy over the preservation of Lake Baikal (Pryde, 1972). The two main reasons why Baikal has been of such great interest are, first, the extraordinary natural features of the lake, and second, the unusual degree to which public opinion played an early role in the controversy, and continues to do so.

Lake Baikal, located in East Siberia just north of the Mongolian border, is perhaps the most remarkable fresh-water lake in the world. Geologically, it lies in a huge graben (a structural depression between two parallel fault systems), and is approximately 700 km long (Figure 5.3). The lake has the distinction of being the most voluminous freshwater body in the world (23,000 cu. km), and the deepest as well (1,620 m). Lake Baikal's main significance is not just its size, however, but rather its biology. In its unusually pure waters can be found over 800 species of plants and about 1,550 types of animal life. More importantly, the majority of these are endemic, making Lake Baikal an object of worldwide scientific importance. Understandably, the threat of its pollution has drawn a great deal of attention both within the Soviet Union and abroad.

Relatively small logging and industrial processing operations have existed around Lake Baikal's shores for several decades. Through the 1960s, their cumulative impact on the lake had not been significant. Then, a greater threat to Lake Baikal arose in the form of two large wood processing plants, proposed to be built at Baikalsk and near the delta of the Selenga River at Selenginsk (Figure 5.3). These new plants would be supplied with timber logged on the steep mountains surrounding the lake, posing the threat of greatly increased soil erosion downslope and into Baikal's waters, which has been estimated at 3 million tons a year (Wolfson, 1988a).

The bulk of attention has been focused on the two new factories. Wood processing plants of this type are inherently polluting, and a debate soon arose over whether the planned purification facilities would be adequate to protect Baikal's exceptionally clean waters. Not unexpectedly, the ministries proposing the plants said they would, but a great many others disagreed. The controversy became widely debated in the Soviet press during the late 1960s, with scientists,

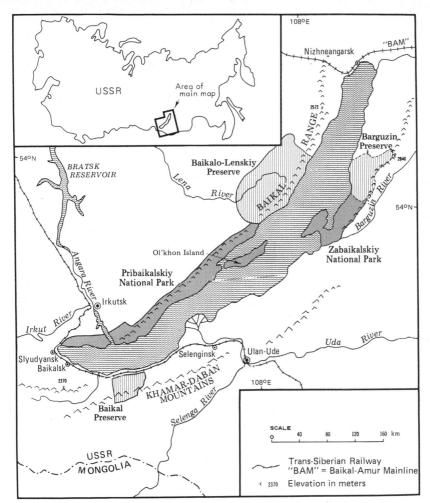

Figure 5.3   Lake Baikal and environs

artists, writers, and many others rising to defend the lake. These exchanges took on significance as the first major environmental controversy in the Soviet Union to be so openly scrutinized, long before the advent of *glasnost'*.

Despite all the publicity, there was never any real possibility that the permits for the plants would be revoked; the best that the opponents could hope for was additional protection for the lake. Their efforts were to a certain degree successful, for the Soviet government did adopt several additional safeguards.

A special resolution on protecting Lake Baikal was passed by the

Council of Ministers in 1969 (and updated in 1971 and 1977), and a commission was established to monitor water quality in the lake. The wood pulp produced at the Baikalsk plant would be sent elsewhere for processing, and if significant pollution was still possible, a pipeline was authorized that could divert the Baikalsk plant's effluvia past the lake and into the Irkut River. Finally logging activities were to be curtailed on the surrounding steep slopes, and logs could no longer be floated loosely on the lake.

Did all this mean that during the 1970s Lake Baikal had been given adequate protection? The ministries in charge of the plants said yes, and insisted the lake would not be damaged (Pryde, 1983b; Rasputin, 1986). But the USSR Academy of Sciences stated in 1977 that the danger of Baikal becoming contaminated had increased, and that the lake was on the verge of irreversible changes (Komarov, 1980). Further, the chief limnologist at the lake has always maintained serious doubts as to whether the measures taken would prove adequate (Galaziy, 1981). Some of the pollutants from the new factories are highly toxic, and even a small volume of effluents, perhaps acceptable in an ordinary water body, might do irreversible harm to the unique biology of Lake Baikal (Wolfson, 1988a). It is known that the Baikalsk plant's treatment facility has been shut down more than once for repairs or improvements, and that its discharges met standards only through the unacceptable expedient of diluting them with lake water before discharging them into the same lake. Also, the Selenginsk plant, opened in 1975, and other polluting factories along the shoreline still operate.

Given the rarity and uniqueness of Lake Baikal's biotic life, it was always clear (at least to scientists) that greater caution should be the watchword. Finally, after more than two decades of debates, studies, and construction, another resolution was passed in April of 1987 that may have finally resolved the issue in favor of the lake. This decree, entitled "On Measures to Ensure the Rational Utilization of the Water Resources of the Lake Baikal Basin in 1987–1995," required a revised general plan for the development of the basin, moved the cellulose operation at Baikalsk to Ust-Ilimsk, ordered conversion of the Baikalsk plant by 1993 to furniture manufacturing, terminated plans for the Irkut River diversion pipeline (which itself had become highly controversial), required a new closed-cycle water system for the Selenginsk plant, and reduced timber cutting along the lake's coastline (*Pravda*, May 10 and July 29, 1987). At the same time, a number of ministry officials were fired or reprimanded.

In addition, during the 1970s and 1980s, a large number of protected areas were created in the Lake Baikal basin (Vorob'yev and Martynov, 1989). The largest of these are shown in Figure 5.3. One of them, the new Pribaikalskiy National Park, takes in some of the most frequented tourist spots on the west side of the lake. These new parks will be described more fully in chapter 9.

Over the years, the public has become increasingly involved. The "Baikal Movement" arose to oppose the diversion pipeline, and more recently a Society for the Defense of Baikal, a Baikal National Front, a Baikal Fund, and other allied groups have been formed (Stewart, 1990).

The 1987 resolution also called for pollution abatement facilities at other enterprises around Lake Baikal, to further ensure that past mistakes would not be repeated. For example, after the start of construction on an apatite processing plant on the Selenga River, which contributes half the water inflow into the lake, it was determined that this plant might pollute both underground and surface waters. As was the case at Baikalsk, it took a determined public outcry before the factory's designers agreed to undertake comprehensive studies of its environmental consequences (*Pravda*, Dec. 28, 1984).

But has the lake really been saved? Although the 1987 reforms are under way, progress on them is lagging considerably behind schedule. And a hundred smaller enterprises around the lake still discharge untreated effluent (Filipchenko, 1989). Thus, the lake's fate appears to remain undecided, and its status is of sufficient concern that three pages of the 1989 environmental statistical handbook were devoted to it (*Okhrana*, 1989, pp. 132–34). By the end of the 1990–95 five-year plan, it should be clearer whether Lake Baikal has indeed been preserved.

Coastal conservation

As in other countries where land and saltwater meet, the coastlines of the Soviet Union have not been immune from pollution problems. Both industrial and municipal effluents have created water quality problems in bays, harbors, coastal seas, and beachfront areas. Even arctic seas are affected; in 1989 it was reported that "most of the rivers flowing into the Barents Sea are catastrophically polluted due to the operation of ore and chemical industries" (Zelikman, 1989).

Offshore oil drilling is not a major activity in the Soviet Union, occurring mainly in the Caspian Sea where output is declining. Minor oil spills and slicks periodically occur. However, exploratory drilling has taken place off Sakhalin Island and in the Azov and Black seas, and more recently in the Baltic and Barents seas (Kelly *et al.*, 1986, p. 119). Anywhere there is oil, accidents occasionally occur. In 1977, a Soviet tanker went aground in Sweden, producing the worst oil spill up to that time in the Baltic Sea. Four years later, a British tanker broke apart in a storm just offshore from the port of Klaipeda, in Lithuania. The spilled boiler fuel from that accident was reported to have formed a huge slick that soon covered about 40 km of sandy beaches to the north, including those near several resort towns. It also penetrated 20 km into the shallow Kurland Lagoon south of Klaipeda. Soviet officials estimated the damage at 600 million rubles (Shabad, 1982). Oil spills from tankers have been a recurring problem in the Black Sea as well (Komarov, 1980, p. 37).

The Baltic coast seems to be a particular problem area. As early as 1976 a special resolution "On measures for stronger protection of the Baltic Sea basin against pollution" was passed by the Council of Ministers. It called for improvements in the control and disposal of such pollutants as municipal sewage, oil products, pesticides, and other by-products of agricultural production. But more than once since then, Latvian beaches have had to be isolated. In 1988 the beach at Jurmala (near Riga) was closed due to severe bacteriological pollution; private car access to the city was by special pass only (*Izvestiya*, July 17). The beach at Parnu in Estonia was closed in 1988 by wastes from nearby slaughterhouses. Cellulose factories and municipal wastes from such cities as Leningrad, Riga and Ventspils have been cited as the major sources of Baltic pollution (Doklad, 1989, p. 59).

To help protect water quality in the Baltic Sea, the Sloka pulp and paper mill at Jurmala was closed on January 1, 1990, despite the resultant loss of jobs. As a consequence, newspaper publication in the Latvian republic had to be temporarily suspended (Litvinova, 1990).

During the same summer, an even worse problem existed at the popular resort city of Sochi, on the Black Sea. Breakdowns of the greatly overloaded city sewage treatment facilities forced a prohibition on swimming more than ten times during the summer and caused a reported 300 cases of gastrointestinal illness (*Izvestiya*, Sept. 11, 1988).

Perhaps the most serious and longstanding coastal problem facing Soviet planners has involved the Sea of Azov. This shallow and biologically rich appendage to the Black Sea in the past has been one of the USSR's most productive fisheries. But starting in the 1970s, there were repeated warnings that the Azov's resources were in danger from pollution and increased salinization, and a special resolution on preventing pollution in the Black and Azov sea basins was enacted in 1976 (*Pravda*, Feb. 4; Komarov, 1980). A dam was proposed to be built across the Kerch Strait to prevent saline Black Sea water from intruding into the Azov, but it quickly came under both economic and environmental criticism and was not built (Mote and ZumBrunnen, 1977).

The primary cause of the Azov's deterioration has been diversions of water from the streams that feed into it, up to 80 percent in the case of the Kuban River. This is rather ironic, given the flooding problem around the Kuban Reservoir noted in the last chapter. This reduction of inflow has increased the sea's temperature and salinity. As a result, there has been a corresponding reduction in fish catches there by 60 to 90 percent. The construction of fish breeding facilities has failed to restore the previous numbers. Similar problems exist at the estuaries of the Dnepr, Dnestr, and other rivers entering the Black Sea, which itself has shown signs of biological deterioration (Rozengurt and Herz, 1981; Tolmazin, 1979).

Pollution from cities, industries, and farms has also been a problem in the Sea of Azov. The 1976 resolution referred to earlier required oil wastes to be isolated, collected, and disposed of without harm to the environment, but by 1988 fertilizer and pesticide residues, and the continuing problem of poorly treated sewage, were being cited as at least equally serious concerns. The combination of polluted run-off, over use of beaches, and heated water caused the entire coast of the Sea of Azov in Donetsk Province to be closed to swimming in the summer of 1988 due to "extremely high pollution levels" (*Izvestiya*, July 23).

An unusual coastal protection problem has arisen in Leningrad, where a huge barrier device is being created to hold back the sea. The problem has been that Leningrad, a low-lying city barely elevated above the Gulf of Finland, has long been subject to damaging flooding from storm-induced tidal surges, called "long waves." The last major inundation occurred in 1924, when water in the harbor rose 3.7 m above sea level. Lesser flooding has occurred several times since then, most recently in 1978, 1982, and 1984. The proposed solution, authorized in 1979, was a 25 km-long barrier dike stretching across the Gulf that would rise 8 m above the sea and block its periodic surges. Six passageways will permit navigation through the structure; massive steel doors will close them off when flooding threatens (Precoda, 1988). The northern half of the dike, from Kotlin Island

(Kronstadt) to the north shore of the Gulf of Finland at Gorskaya, was completed in 1984 (Figure 5.2). The southern portion will not be completed until after 1990.

But such a solution is not without its own problems. Complaints were expressed in the press in the late 1980s that the dike was hampering normal water circulation in the diked-off portion of the Gulf and turning it into a stagnant settling basin for Leningrad's effluvia (and that coming down the Neva from Lake Ladoga). Associated effects are algae blooms, noxious odors, and a deleterious effect on fishing and coastal recreational opportunities. Other predictable problems are ice build-up along the dike in winter and rapid sedimentation of the shallow basin behind the dike, both of which will adversely affect Leningrad's maritime economy (Precoda, 1988). The environmental effects of the dike appear to have been astonishingly poorly analyzed, and it is derided by many Leningraders. Despite all this, a special commission has recommended it be completed, but with design modifications and permanent monitoring (*Leningradskaya pravda*, Apr. 21, 1989, p. 2).

Policy implications

Soviet public health officials fully appreciate the necessity of providing adequate water purification facilities. Yet it is apparent that serious institutional and organizational problems remain. These problems are rarely technical in nature, they are almost always found in the realm of policy implementation. The 1990s agenda for improving water quality in the USSR can be summarized in three words: priorities, funding, and enforcement.

As noted earlier, the USSR uses a system of Maximum Permissible Concentrations (MPCs). As noted earlier, these in general are fairly strict, at least on paper. The problem is enforcement, which is frequently lax; oil concentrations, for example, are often 20 to 60 times the MPC level (Komarov, 1980). The problem of "narrow departmentalism" – the tendency of one ministry to look after its own interests to the exclusion of broader planning and public concerns – has also shown great resistance to reform (ZumBrunnen, 1984; Jancar, 1987; Ziegler, 1987).

There is no question that on paper, at least, water quality is a high priority; the need is to translate that paper priority into continuous action within the ministerial bureaucracies. But both ministerial and factory managers know that bonuses are awarded for meeting production quotas and opening new plants on time, not for delaying an opening because the treatment facility was not quite finished. Greater incentives to meet environmental goals, in the form of either a carrot or a stick, need to be put in place. The new "ecological program" for the USSR being considered by the Supreme Soviet in 1990 will hopefully contain the provisions to accomplish this.

There is probably no country on earth in which funding for water pollution control is as great as it should be. Despite the best of official intentions, the USSR lacks adequate funding both for new purification facilities, as well as for maintaining and expanding older ones. In addition, even those funds that are allocated may not be used; *Pravda* pointed out that from 1975 to 1985 authorized pollution-control funds were underutilized by 15 percent (Poletayev, 1987). Whether for lack of funds or other reasons, only 58 percent of the water purification facilities planned to be built in 1984 by the Ministry of Industrial Construction were completed (Khrenov, 1985). In 1987 *Izvestiya* further noted that during the 11th five-year plan the ferrous metallurgy industry met its goals for reducing untreated wastewater releases by only about half. In the Soviet ministerial system, it has traditionally been very expedient to relegate water quality to a lower priority than fulfilling industrial norms.

It is necessary to institutionalize an effective program to ensure that water quality directives will be fulfilled. Four areas where improvements are needed can be identified: first, purification facilities must be in place when new factories are opened; second, expanded factories must have correspondingly enlarged water treatment capabilities; third; adequate money must be made available for ongoing maintenance and repairs, and fourth, new technologies must be more quickly incorporated into the construction agenda.

It is not that enforcement efforts do not exist, nor that they can not be effective. The closing of the Priozersk mill has already been noted. However, it probably should have been shut down years earlier, and the four different directives concerning the protection of Lake Baikal (as well as the recent Filipchenko article) suggest foot-dragging, as well. One problem may be understaffing; Ziegler notes that the Sanitary Epidemiological Service of the Ministry of Health has only a little over a hundred stations throughout the country, an average of fewer than one per each major administrative unit in the USSR (Ziegler, 1987, pp. 116–17). It is also generally agreed that the present fine system is ineffective, the relatively small fines being less onerous than ensuring that treatment facilities are built in a timely manner.

Nor are violators of water quality laws rigorously prosecuted; Ziegler cites a figure of only 25 percent of violators actually having charges filed against them (p. 100).

There is a growing consensus that the Soviet Union's water quality situation is critical and that major changes are needed. In the words of one scientist, the laboratory director of the Academy of Sciences' Institute of Inland Waters:

We are now standing on the brink, and if we cross it, a nationwide ecological catastrophe is inevitable. We must recognize that the path we have taken in the past 30 years in search of a "cheap" solution to the problem of protecting bodies of water from pollution is a dead end, and the premise that the problem can be solved through the self-purifying ability of bodies of water is fundamentally incorrect and lacks all scientific basis. We need a radical reassessment of the role and importance of water resources in the life of society, and a primary emphasis on the life-supporting function that bodies of water perform for all living things . . . (Lukyanenko, 1989)

Change may be in the air, or at least an understanding of the need for it. The head of the USSR Environmental Protection Committee was quoted in 1988 as saying "we are sinking in instructions, directives, and declarations . . . It is a crisis situation; we need a 'new ecological thinking.' We must eliminate red tape and bureaucracy" (Reuters, Aug. 19). But "bureaucracy-bashing" alone will not make the problem go away. New, unresolved water quality problems arise as fast as previously targeted ones are corrected. Policy directives must emphasize enforcement, and funding, more than in the past. One source estimates 36 billion rubles a year are needed to fully comply with normative standards (Sebastian). Perhaps the new Environmental Protection Committee (Goskompriroda), and the forthcoming State Ecology Program will have the muscle to overcome ministerial inertia and implement the needs that have been outlined above; if so, it will be a signal victory for *perestroika* on the environmental front.

6 Controlling toxic and urban wastes

No clear distinguishing line can be drawn between the concepts of "air and water pollutants" and "toxic wastes." There are two primary reasons for this. The first is that the term "toxic waste" is not well defined: at what concentration does a particular compound become toxic, and to whom, human beings or smaller, more sensitive (or perhaps more tolerant) organisms? The second reason is that modern society manages to put its chemical residuals into all types of environments where they are often transformed; sulfur and nitrogen compounds, for example, can wind up in air, water, or soils, with varying degrees of toxicity depending on their form and location.

Still, there is merit in devoting a specific section of a book on environmental management to the subject of "toxic wastes," both to underscore their critical importance, and to cover several categories of health threats that are often left out of traditional discussions of air and water pollution. These threats include certain types of industrial chemicals, harmful metals, low-level radioactive materials, pesticides, household toxic wastes, and various residuals from all forms of waste disposal (including the familiar municipal "trash dump").

A highly industrialized society such as the Soviet Union cannot be free from these unwanted chemical intrusions. They are the dark side of "progress." Speaking only of agricultural lands, one Soviet author reported that "around 1975–76, when special soil laboratories were set up, it was found that the country's fields and pasturelands were contaminated with 150 kinds of pesticides, poisonous chemicals, and trace elements" (Komarov, 1980, p. 47).

In recent years, the Soviet chemical industry has become much better known (Sagers and Shabad, 1990). In the Soviet Union, chemical and other industrial enterprises are responsible for the proper disposition of their waste products, including toxic ones. Placing the onus for safe disposal on the producing firm can be an effective control strategy, if there is adequate oversight of the firm's compliance with these regulations. Unfortunately, there is considerable evidence that this oversight function in the Soviet Union is often either lax or ineffective, as it is elsewhere.

Industrial residuals

Industrial processes inevitably use, create, and are forced to dispose of a wide variety of potentially hazardous materials. Some of these, by any definition, are clearly toxic to human beings. In some cases, these materials may escape (or be discharged) into the environment through either natural or man-made water conduits. In many other cases, these materials are isolated in special containers for safe disposal elsewhere.

One of the most widespread of the dangerous industrial chemicals is a compound called polychlorinated biphenyls, usually shortened to the acronym PCBs. These are a liquid plasticizing agent, commonly used in the manufacture of a variety of products such as certain plastics, paints, inks, adhesives, and phonograph records. They are also an excellent non-conducting insulator, and thus are widely used in electrical transformers such as those seen on telephone poles. Their main drawback is their extreme stability; no environmental agent, including heat, is readily able to destroy them.

Concerns about PCBs first arose in Europe and the United States in the late 1960s, when it was discovered that their breakdown products strongly resembled those of DDT. They were soon identified as a problem in the Soviet Union as well, but by Swedish scientists rather than Soviet, who discovered them in 1974 flowing out of the Neman River and into the Baltic Sea. Komarov relates how at first there seemed to be no industrial sources that could account for their presence, but that an internal check showed them to be present in virtually every water body in the country. The cause was later acknowledged to be the military, to whom they had been delivered since the 1950s from a manufacturing plant near Gorkiy (Komarov, 1980, p. 33).

Yet even as recently as 1988 the problem of controlling these chlorinated chemicals appeared to remain: *Pravda* quoted a USSR institute director as saying that "of the major users of chlorine and chlorine products, only our country has not placed the problem of substances such as polychlorinated polycyclics (which includes PCBs) within the framework of a national program and is not participating in international collaboration on this problem" (Fokin, 1988).

How might such long-lived chemicals be disposed of? There are no easy answers, but one possible technique is very high-temperature incineration, which can be conducted either on land or at sea. The International Union for Conservation of Nature (IUCN) reported in its March 1988 *Bulletin* that trial incinerations of this type had succeeded in destroying PCBs by factors of from 99.997 to 99.99999 percent. Tests on a large number of other hazardous chemicals had produced similar results. Such burning at sea is opposed by groups who say that even these impressive-appearing burn-up percentages allow significant amounts of harmful residues to enter the marine ecosystem when large volumes are incinerated. Although controversial, at present no other suitable process for the destruction of existing stocks of these types of chemicals is available. To date, the Soviet Union apparently has not experimented in any major way with this technique.

Another possible approach to the permanent storage of hazardous wastes is placing them in underground cavities created by nuclear explosions. At least one report has suggested that this has been carried out in the USSR for disposing of chemical wastes such as oil field brines (Browne, 1983).

A category of industrial residuals that is of great concern is toxic metals. Although these have sometimes been referred to as "heavy metals" since they include lead and mercury, some of the elements of most concern are quite light (e.g., beryllium). All are potentially injurious to human health. Certain of these metals are also suspected of causing observed tumors in fish taken from contaminated marine environments.

Unfortunately, many problem areas involving toxic metals have been reported in the Soviet Union. These frequently occur in or around the towns where the ores are smelted, or where they are used in industrial operations. A conference in Kalinin in 1976 identified lead concentrations of five to six times the Maximum Permitted Concentration (MPC) from several regions; in the city of Leninogorsk in Kazakhstan the air contains on average 30–40 MPCs of lead, and has risen at times to 440 MPCs. In nearby Ust-Kamenogorsk, lead concentrations have reached 14 MPCs. Other Soviet cities in which toxics-emitting metal smelters have operated for decades include Chimkent, Ordzhonikidze, Almalyk, and Norilsk. The pollution problems in Norilsk have been reported upon and found to be severe (Bond, 1984).

Mercury is one of the most hazardous of metals. Airborn mercury levels in Sterlitamak are at 10 MPC, and in the Kazakh city of Temir-Tau they are at 60 MPCs. At the Solikamsk caustic soda plant, mercury is released at about three times the permissible standard, but

Figure 6.1   Moldavia and western Ukraine *Source: Atlas SSSR,* 1983

this is about forty times the figure for similar production in Japan and Finland (Komarov, 1980). In Smolensk, Klin, and Saransk, factories that manufacture light bulbs have routinely deposited mercury-containing wastes in city dumps or abandoned quarries (Pipia, 1989).

Less familiar metals can contaminate the environment as well. In one recent case, high doses of thallium were recorded in the Ukrainian city of Chernovtsy, near the Romanian border (Figure 6.1). Soviet news reports indicated that exposure of the population to this toxic element caused at least 127 children to lose their hair. It was initially hypothesized that the thallium had originated outside the USSR, and that it had been deposited by a rainstorm (*Pravda*, Nov. 17, 1988), but later it was acknowledged that it had probably originated in industrial wastes buried within the city limits (*Izvestiya*, Nov. 19, 1988). Ultimately, 90 percent of all pre-school-age children in Chernovtsy were evacuated. Elsewhere, cadmium was encountered in an area near Moscow in concentrations 100 times the norm, and boron has been a problem in the city of Tatarsk (Komarov, 1980). No reports have been seen concerning the metal selenium, which has damaged wildlife refuges in California. TASS reported an explosion of toxic beryllium oxide gas in Ust-Kamenogorsk in September, 1990.

In most of the examples cited above, a probable cause would appear to be a weak compliance with regulations and norms. The most likely explanation for this is a shortage of capital for producing equipment to capture and recycle these metallic residuals. In theory, there should be a strong incentive for doing this, as most of them represent resources that are needed in the economy. Change-over costs may explain the slow pace of banning leaded gasolines in the USSR; in 1989 only 30 percent of Soviet gasoline was unleaded, although a complete phaseout is reportedly planned by 1997.

Although not an element, asbestos is a potentially lethal mineral product; its fibers, if inhaled, can severely injure the lungs. Its dangers are well known, but it seems to have received little discussion in the Soviet press. Asbestos, mined in the Ural Mountains and the Tuva republic, has been widely used in the Soviet economy. Indeed, in the 1960s the USSR moved into first place in the world in asbestos production (Shabad, 1969, p. 66). Given the troubles that it has caused in the United States, it would be surprising if at least some local workplace or building problems involving asbestos had not occurred somewhere in the Soviet Union.

Environmentally damaging accidents at industrial sites can have a variety of causes. *Izvestiya* reported the collapse of an industrial waste dam on a tributary of the Dnestr River on September 15, 1983 (Zakharko, 1983). The dam, near the town of Drogobych in the western Ukraine, was used to hold highly saline wastewaters from a plant producing potassium fertilizers. The resulting uncontrolled flood produced a wall of water 6 m high which tore away 400 m of railway tracks. The contaminated floodwaters polluted first the tributary and then the Dnestr River itself for over 500 km downstream (see Figure 6.1). The salty flow deprived many cities and industries in both the Ukrainian and Moldavian republics of their primary source of water supply. In addition, it resulted in significant fish kills along the affected portion of the rivers. It took months of natural flushing to restore the river to its normal condition, but much of the salty wastes settled on the bottom of a downstream reservoir, where they still pose a potential problem. A special commission was created to investigate the causes of the dam failure.

Pipelines and storage tanks can also burst; a pipeline break in February of 1989 in West Siberia was burned off without fatalities. But tragedy occurred four months later when a passing train detonated a huge pool of escaped gas from a liquified natural gas pipeline near Ufa that killed hundreds of the train's passengers. In Jonava, Lithuania, a large storage tank containing ammonia exploded in 1989 at an industrial site, killing at least four persons and injuring dozens. The poisonous cloud of ammonia gas necessitated evacuating 30,000 people before it dissipated (*Izvestiya*, Mar. 21, 1989, p. 8).

In the USSR, as in the US and Europe, barges of municipal wastes sometimes become "unwanted orphans" in search of a port that will accept them. This happened to the Soviet ship *Petersburg*, which in 1989 was prevented from being unloaded in the town of Ust Dunaisk on the Black Sea, due to protests from residents (Wolfson, 1989b).

Radioactive wastes

Although radioactive waste disposal was mentioned in chapter 3, that discussion was limited to wastes produced by the commercial nuclear reactor program. There are many other sources of radioactive wastes. Specific divisions within many research institutes, enterprises, universities, and hospitals use a wide variety of radionuclides in their everyday operations. Low-level wastes from research reactors at scientific institutes and universities are usually handled in a similar manner to those from commercial reactors. Other low-level wastes are, in theory, closely regulated; Maximum Permissible Concentrations have been established for 723 radioactive substances (Jancar, 1987, p. 112).

Although today specific measures are prescribed in the USSR for the safe disposal of these low-level wastes at state sanitary-epidemiological stations, this was not always the case. In the 1940s and 1950s, radioactive wastes were buried at sites on the (then) outskirts of Moscow. Today, Moscow has spread out into these regions, but adequate records of the exact location of these sites have been lost. A newspaper called this "negligence pure and simple" (*Moscow News*, 1989, no. 18, p. 8). As a result, radiation surveys must now be done in these suburbs in areas of new construction.

But current regulations are seemingly not always followed, either. The same issue of *Moscow News* also noted that drums of radioactive wastes had been found by schoolchildren on an outing near the city of Khabarovsk. A local enterprise that used radon admitted to the improper dumping, which had been going on since 1961. Radioactive wastes were found on the grounds of a reinforced concrete plant in Kirovograd (Schoenfeld, 1989). An ampule with a discarded medical instrument containing the radioactive isotope cesium-137 was found in a school flower bed in Krasnoyarsk (*Pravda*, Oct. 22, 1987, p. 6). Although it did not raise ambient radiation above the normal background level, the potential danger it represented was clear. Other cities with radioactives waste problems include Irkutsk, Kirovograd, Tbilisi, and Podolsk (*Izvestiya*, July 13, 1990, p. 7).

By far the most serious accident in the USSR involving radioactive wastes, one that probably released the greatest amount of radiation of any accident prior to Chernobyl, was the Kyshtym incident. This accident occurred at a secret nuclear weapons processing plant in the Urals in September of 1957, at a site in between the towns of Kasli and Kyshtym.

For over thirty years, the Kyshtym accident was never officially acknowledged in the Soviet Union, despite overwhelming evidence that it took place (Medvedev, 1979). Finally, in 1989, the Soviet Union admitted that it had happened, stating that 2 million curies of radioactive elements were released from an exploded waste tank, and were deposited over an area 105 km long and 8–9 km wide (*Tass*, June 16, 1989; *Science*, 244, June 23, 1989, p. 1435). Over 10,000 persons were evacuated, and a large area of surrounding countryside in the drainage of the Techa River became heavily contaminated (Figure 6.2), although apparently the nearby large city of Chelyabinsk was not endangered. For years, the area was placed off limits to all human contact, and parts of it are still restricted today. One lake is reportedly so contaminated that it must be completely filled in with concrete (*New York Times* news service, July 10, 1989). The Soviet Union indicated an intention to decommission and dismantle some of the military reactors at this site, which may well be at the end of their useful life anyway, in 1989 and 1990.

One other radioactive decay problem should be mentioned, and that is radon. Radon gas is a naturally occurring low-level radioactive

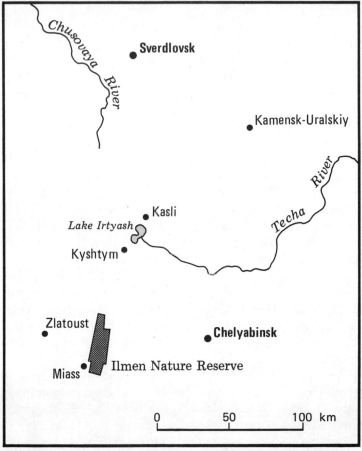

Figure 6.2   The Kyshtym–Chelyabinsk region *Source: Atlas SSSR*, 1983

emission given off from certain ancient rock formations. Radon concentrations can be substantial in tightly sealed homes built in areas where it is present; this has been a problem in certain areas in the eastern United States. It is not known what areas, if any, might be potentially affected in the Soviet Union by radon gas.

Since toxic industrial (and military) wastes are inevitable, their safe handling is essential. This involves safety in their manufacture, in their use, in their storage, in their transport, and in their ultimate disposal. To better achieve these goals in the Soviet Union, the following recommendations have recently been made. First, the USSR State Committee for Science and Technology should formulate a nationwide program for protecting the public and the environment from hazardous materials. Second, a scientific center for studying the safe use of hazardous materials should be established under the State Committee for Environmental Protection (Goskompriroda). And third, a separate analytical center should be established for systematically monitoring on a nationwide basis hazardous substances in the environment (Fokin, 1988). Also recommended was the publication of a journal (the name "The Chemosphere" was suggested) that could chronicle problems and progress in the safe use of hazardous materials.

Toxic wastes, of course, are a worldwide problem. In reflection of this, an international agreement was signed in the spring of 1989 to control the export of hazardous wastes. African nations particularly wanted this treaty, because of a fear that the more industrialized nations might want to "dump" toxic wastes in developing countries. The treaty will ban the export of toxic wastes to countries not prepared to handle them safely.

Pesticides

Pesticides pose problems in all parts of the world, but tend to be most serious in countries that have a strong agro-chemical industry, and agricultural and forestry sectors that are characterized by monocultures (large fields of a single crop) and a goal of high labor productivity. This is descriptive of both the United States and the USSR.

The Soviet Union has placed a strong emphasis over the past three decades on what it calls the "chemicalization" of agriculture, with pesticide use growing eight times from 1960 to 1977, and still increasing today (Vasil'yev, 1983; Table 6.1). In 1986, pesticides were used on 87 percent of all cultivated land in the USSR, at an average of 1.9 kg/ha (Yablokov, 1988). This is in response to an estimated 90 million tons of agricultural crops that are lost to diseases and pests annually (*Pravda*, Jan. 6, 1981, p. 3). Predictably, this "chemicalization" has led to wildlife depletions, concerned pleas by scientists, and a certain amount of effort to institute better pest control procedures.

As would be expected, a wide variety of pesticides (including insecticides, herbicides, fungicides, rodenticides, etc.) are used in the USSR. The translation journal *Entomological Review* occasionally publishes in English articles dealing with both Soviet chemical and biological control methods. A review of Soviet pest control practices and problems appeared in the early 1970s, prompted by a large amount of material that was appearing at the time on Soviet pesticide

Table 6.1   *Pesticide use in Soviet agriculture (delivery of pesticides to the agricultural sector, in 1,000 metric tons)*

Category	Average per year		
	1976–80	1980–85	1988
Insecticides	73.5	86.3	68.0
Fungicides	194.3	238.5	242.5
Poisons	9.2	7.3	6.8
Herbicides	114.5	148.9	155.6
Defoliants and dessicants	34.8	43.2	43.8

Source: Display in the land use pavilion at the Exhibit of Economic Achievements, Moscow, May 1989

practices (Pryde, 1971). Subsequently, the incidence of articles in the Soviet press dealing with pesticides use greatly dwindled during the later 1970s and early 1980s.

Nonetheless, it is known that pesticides have been frequently misused and overused in the USSR. The 1971 review cited several articles from the Soviet press that described the misapplication of pesticides, the appearance of residues on foodstuffs, and specific cases involving accidental kills of deer, elk, marten, rabbits, squirrels, foxes, grouse, partridges, bustards, snipe, ducks and geese, as well as several species of song birds. A more recent report stated that "pesticide-treated Kuban rice fields yielded the state about 1.5 billion rubles in profit, but caused more than 2 billion rubles in damage to fisheries in the Azov basin" (Yablokov, 1988).

One early incident involved the use of zinc phosphide in Siberia to control susliks (a small rodent); following its application there were recovered 385 dead geese and ducks, but only 25 susliks (Yeliseyev, 1966). Here, as often happens, history repeated itself, for in 1979 came reports that another misapplication of zinc phosphide killed 169 geese at a collective farm near Rostov (*Pravda*, Dec. 6, 1979). Part of the problem may be imprecise areal applications; one official stated that up to 60 percent of pesticides sprayed from aircraft miss their target area (Wolfson, 1989a).

The residues of long-lived chlorinated hydrocarbon pesticides, such as DDT, cannot be broken down biologically, and thus can build up to significant concentrations that remain in the environment for many years. This problem is cited by Komarov, who notes that in the Uzbek republic, where 100,000 tons of pesticides, defoliants, and other

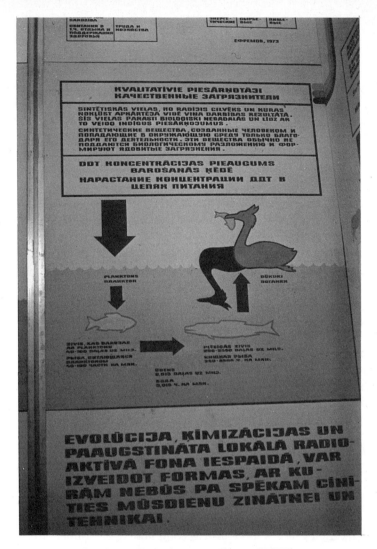

Figure 6.3   Display in Riga showing effects of DDT

chemicals are applied annually, DDT residues were first recorded in the USSR in excess of the maximum permissible concentration. DDT was outlawed in the Soviet Union twenty years ago, but *Izvestiya* reported in 1988 that breaches of this ban may be occurring:

Azerbaidzhan leads in the application of pesticides per hectare. It exceeded acceptable sanitation standards by a factor of 10 on cotton and vegetable fields, and by a factor of nearly 100 in vineyards. The highly toxic pesticide DDT, officially banned in 1970, has been used in agriculture in various years since then (*CDSP*, 1988, 40, no. 36, p. 20)

In addition to Azerbaidzhan, the 1989 State of the Soviet Environment report cites Armenia, Moldavia, Uzbekistan, and portions of the North Kazakh and Novosibirsk provinces as areas where DDT concentrations have exceeded the permitted norms by 2–200 times (Doklad . . ., 1989, p. 83). How and why the more recent applications of DDT took place was not explained, but its harmful effects are clearly understood (Figure 6.3). Dexochlorane (possibly related to the banned American pesticide Kepone) is another troublesome chlorinated hydrocarbon. Its residues have been reported as exceeding norms by thousands of times in the Gulf of Taganrog (*Soviet Geography*, 28, Oct. 1987, p. 610).

A more recent banning involves the Soviet herbicide "butifos," which is commonly used to defoliate cotton fields in republics such as Uzbekistan. Articles appeared for almost a year in the journal *Literaturnaya gazeta* pointing out adverse effects from the widespread use of this defoliant. The initial response from the ministries was to defend it. Finally, after more public complaints, a thorough investigation was conducted, and the Ministry of Public Health acknowledged that butifos was being applied contrary to regulations and in unsafe ways, and consequently was contaminating food products as well as water and air quality, poisoning farm workers and crop sprayers, and lowering public health indices in the areas where it was most heavily used. It was eventually banned by the Ministry of Public Health in March of 1987 ("Toxicosis," 1987). However, as with DDT, there appear to be "exceptions" in the enforcement of the ban (Wolfson, 1989a).

Besides butifos, other herbicides, such as the commonly used 2,4-D, are also widely employed for weed and brush control in Soviet forestry and agriculture. A 1977 article complained of double-strength concentrations of herbicides being used in Karelia to kill young birch trees that were invading commercial pine stands, and quoted local forestry officials as saying that forested areas sprayed with chemicals should be curtailed by two-thirds (*Pravda*, Jan. 31, 1977, p. 4). As with insecticides and rodenticides, these misapplications of herbicides have also resulted in kills of wildlife.

Discussions concerning the possible mutagenic effects of certain types of herbicides (those containing the impurity dioxin) have not been encountered in the Soviet press. One article, reviewing the state of the Volga River, did state that "the high level of toxins from chemical weedkillers and pesticides in reservoirs from which drinking water is drawn is particularly alarming" ("Hands in the Volga," 1988).

The "chemicalization of agriculture" also involves the heavy use of fertilizers which, if misused, can likewise pose threats to drinking water supplies (Shchepotkin, 1987). The possibility further exists of a linkage between regions of excessive use of chemicals in agriculture and high rates of infant mortality (Wolfson, 1989a).

Soviet scientists have been as eager as their western counterparts to substitute effective biological methods of pest control for chemical spraying (Yablokov, 1988). Biological methods involve predator insects, and techniques that disrupt the reproductive or developmental cycles of insects. Considerable research on biological controls has been carried out at institutes in Moldavia and elsewhere, and numerous field tests and commercial applications have been performed, using the common *trichogramma* wasp and other predators. In 1975 it was reported that a factory for the production of a million *trichogramma* wasps a day had been constructed in Voronezh (*Pravda*, Feb. 25). Fourteen other species of insects are also used in the Soviet biological control effort (Yablokov, 1988).

According to press reports, the acreage on which biological or integrated pest control procedures were applied increased from 200,000 ha in 1960 to 20 million ha in 1983 (*Izvestiya*, May 23, 1984, p. 2). The author observed a figure for 1988 of 26.7 million ha at an agricultural exhibit in Moscow. Officials have informally indicated a goal of doubling this area during the 1990s, but this would still be a very small percentage of all agricultural land. The *Izvestiya* article did not provide information on what specific techniques were involved or their effectiveness, but the Yablokov article claimed good results with biological controls.

The various reports on pesticide use (and mis-use) in the Soviet Union that have appeared over the past twenty-five years have included numerous calls for tighter control over the application of these chemicals. In response, new regulations to better govern pesticide applications appeared in both 1979 and 1984. Acceptable pesticides for agricultural use are registered by the State Commission on Chemical Agents for Combating Agricultural Pests. Maximum permissible concentrations for residues on food have been established for 100 pesticides (Jancar, 1987, p. 112). The appearance of numerous articles supporting wide use of biological methods shows that there is concern about reducing dependence on chemical sprayings, and their resulting effects on non-target species. All this notwithstanding, the guidelines for the 12th (1986–90) five-year plan called for the production of herbicides and pesticides to be "substantially increased."

Municipal solid waste disposal

In the past, the disposal of municipal wastes has not been generally classified as one of the more pressing of environmental problems. More recently, however, this subject has drawn increasing attention, for several reasons.

First, mushrooming volumes of solid wastes from ever-expanding cities eventually create serious locational problems with regard to trash disposal. These volumes reach an estimated 27 million tons annually in the cities of the Russian republic alone (Pipia, 1989; Doklad, 1989, p. 77). Second, it is becoming more widely appreciated that municipal solid wastes consist not just of benign paper products and table scraps, but also contain many products that are legitimately classified as hazardous wastes. It is true that many of these products, such as paints, household pesticides, and cleaning chemicals, are probably much less used by Soviet families than by those in Europe or America, but chemicals of this type are certainly used by urban administrative bodies and urban industries, and it is likely that at least some of these find their way to municipal disposal facilities. If such wastes are burned, they become transformed, but generally represent an equally dangerous hazard. Third, as many raw materials become scarce, municipal wastes are increasingly seen as a potential source of economically valuable recycled materials.

The cheapest method of disposing of municipal wastes is burying them in landfill sites. In the 1980s landfills were the dominant method of municipal waste disposal in the USSR, accounting for 97 percent of discarded urban trash in 1988 (Doklad, 1989, p. 77; Raznoshchik and Lobov, 1979). Permits for Soviet landfills, of which about 6,000 exist in the USSR, are issued by Goskompriroda (Peterson, 1990). A landfill site visited in Leningrad in 1988 was not covered over by earth on a daily basis (Peterson and Pickard, 1988); this is a standard practice in the US and elsewhere to minimize potential health threats. Recovery of landfill gas has been carried out only on an experimental basis.

A procedure used in several Soviet cities to dispose of urban trash is incineration. The danger of the improper burning of municipal wastes was demonstrated at a waste dump near the city of Kursk. Here, an incinerator was burning not just urban "garbage" but also wastes from

battery and rubber factories. The smoke from the burning was identified in 1973 as causing significant damage to a section of the nearby
Central Chernozem *zapovednik* (nature preserve), including even its
soils (see Figure 13.3). Complaints from the preserve director almost
resulted in closing the dump, but as of 1980 it was still operating
(Komarov, 1980, p. 56). Air pollution from uncontrolled burning at a
municipal dump has also been reported from the city of Togliatti
(Shalgunov, 1984).

Several articles appeared in the 1970s that discussed municipal
waste disposal practices for the city of Moscow. All of the problems
relating to solid waste disposal that were noted at the start of this
section were identified in the context of Moscow in the early 1970s.
The national capital produced at that time 1.3 million tons of municipal
wastes per year – a far lower *per capita* amount than in America or even
European cities (*Izvestiya*, Jan. 6, 1973). However, the amount doubled
between 1966 and 1977. Both a suburban dump site and incineration
plants were being used to handle the wastes. At this time, an
advanced waste treatment plant, with a capacity of a half million cu.
meters of garbage a year, was built in Moscow. The plant features
recycling of ferrous metal, the production of an organic soil conditioner (35,000 tons/year), and the export of heat to nearby industrial
buildings (*Moscow News*, 1974, p. 37).

A similar plant was built for the Leningrad area, although it can
handle only 25 percent of the city's wastes; two more plants are
proposed to be built in the 1990s. Facilities to turn trash (and sewage
sludge) into an organic soil conditioner also exist in Minsk and
Tashkent. The Leningrad operation includes an experimental pyroly-
sis plant (one that burns wastes at high temperatures in the absence of
oxygen, thereby minimizing pollutants) which produces charcoal,
biogas and tar (Raznoshchik and Lobov, 1979; Peterson and Pickard,
1988). In all, there are eight plants in the USSR that turn municipal
wastes into compost and soil conditioners. Plants that burn trash to
provide district heating exist in Odessa, Tbilisi, Vladivostok, and other
cities.

An American delegation had the opportunity in 1988 to visit Soviet
municipal waste disposal facilities in Moscow, Minsk, and Leningrad.
The plants mentioned above in these three cities were inspected, and
information was received on new and proposed waste treatment
facilities in Orel and elsewhere (Peterson and Pickard, 1988). The Orel
plant will produce a refuse-derived fuel. At least a dozen waste-to-
energy plants currently exist in the Soviet Union (Table 6.2). The

Table 6.2   *Energy recovery facilities in the USSR*

City	Republic	Facility size (tons per day)	Year of commission
Dnepropetrovsk	Ukraine	1,584	1987
Kharkov	Ukraine	1,188	1982
Kiev	Ukraine	1,584	1988
Moscow	Russia	440	1974
Moscow	Russia	1,200	1983
Murmansk	Russia	792	1984
Pyatigorsk	Russia	1,188	1982
Rostov	Russia	1,188	1983
Saratov	Russia	1,188	1986
Sochi	Russia	792	1982
Vladimir[a]	Russia	200	1982
Yalta	Ukraine	1,188	1981

Note: [a]Soviet developed mass burn plant
Source: Peterson and Pickard, 1988

primary institute in the USSR that trains specialists and carries out research in urban sanitation techniques is the Department of Municipal Sanitation of the Panfilov Academy of Municipal Economy, which is under the Ministry of Housing and Municipal Economy.

Planning for the future: recycling

Recycling is widely considered to be the wave of the future in handling a great many types of both solid and liquid wastes, and is currently being strongly advocated in the Soviet Union. Recycling not only reduces the need for landfill space, but also somewhat decreases the need for extracting new raw materials from the earth, and in the process reduces energy demands. A related concept is source reduction, which implies reducing the input of toxic compounds into the economy in the first place.

The construction in Moscow, Minsk, Leningrad, and other cities of urban waste processing plants that extract and recycle ferrous metals and certain other resources was noted above. These plants are commendable and should be standard practice in all large cities, but they are only part of the overall solution to the municipal wastes problem. Recycling prior to the act of waste collection is even more desirable.

Moscow has had programs for recycling waste paper since at least the early 1970s, when over 400 special paper recycling centers were set up in all of the city's boroughs. An investigative report, however, noted that only 1.5 tons per day of scrap paper was being collected, not very much from a city of 8 million people (*Pravda* alone distributes 300 tons of newspapers daily in Moscow). Overall in the USSR, less than 20 percent of paper products were being recycled in the mid-1970s (Tsekov, 1974); a decade later this may have risen to a little over 30 percent (Wolfson, 1988b). In the late 1980s, around 2 million metric tons of paper products and textiles were being recycled annually in the Soviet Union (Peterson and Pickard, 1988); this figure was not expressed as a percentage of the total.

In addition to Moscow, similar recycling efforts were being conducted in over twenty other cities. In an interesting experiment to increase the tempo of recycling, people who brought in used newspapers to the recycling centers were given high quality literary books in exchange (Peterson and Pickard, 1988). This resulted in a 25 to 30 percent increase in total paper collections in Moscow and Leningrad, but even so less than a third of all paper was being turned in (Ivchenko, 1975). In addition to paper, many cities also have collection points for recycling glass bottles. Aluminum cans are not a widely used container product in the Soviet Union.

The concept of recycling is appropriate not only for household products such as newspapers and glass jars, but also in industrial operations as well. Since the early 1970s, the Soviet Union has placed considerable emphasis on the creation of "low-waste" and "non-waste" industrial technologies. This concept appears to involve two main considerations: the full use of all beneficial resources in the extraction of mineral and timber resources, and the maximum recycling back into the economy of residual products from manufacturing processes. Although this concept has long been state planning policy, it seems to be lagging in implementation. In 1984, the official in charge of the program was reported to have said the "the first and only chemical plant planned as a non-waste enterprise (the Pervomaisk combine in the Ukraine) had not achieved its aims in the course of a decade" (Wolfson, 1988b). Earlier, it had been noted that this plant was meeting its "non-waste" mandate not by recycling its residual products into the economy, but by burying them in the ground (Komarov, 1980, p. 92).

Hazardous wastes from industry can be handled in three ways: (1) source reduction (find ways to use fewer hazardous materials as industrial inputs); (2) recycle to the maximum extent possible those that must be used; and (3) safely eliminate, by incineration or other means, those that cannot be recycled. If (1) and (2) are faithfully pursued, well under 10 percent of the input chemicals should need incineration or other forms of ultimate disposal.

One proposed Soviet effort along these lines has been to use ash and slag from coal-burning power stations to build roads and other structures. Yet Wolfson (1988b) cites an overall USSR figure of only 11 percent for the reuse of this ubiquitous power plant residual in the economy.

Another example helps to identify some of the causes of the problem. A chemical plant in the Krasnoyarsk region was supposed to recycle highly harmful fluorine gases and sell them to other enterprises that needed this chemical. However, it took five years to accomplish this, during which time a large area of the surrounding forest was destroyed by the fumes. The problem was that neither the chemical plant nor the separate agency within the Ministry of the Chemical Industry that was in charge of fluorine production wanted to bear the cost of the recovery facility (Komarov, 1980, p. 20).

The primary causes of these problems, ones cited frequently by Soviet commentators, were (1) the longstanding dilemma of "departmentalism" (i.e., lack of inter-ministerial cooperation), and (2) the absence of adequate economic incentives for anyone to capture and recycle usable by-products, such as the fluorine, as quickly as possible. As another example, a water recycling facility at the Taganrog steel mill went under construction in 1975 and was due for completion in 1978, but by 1989 only 57 percent of the allotted capital had been used (Morgun, 1989).

These problems will surface frequently throughout the course of this book. Certainly one of the major challenges of *perestroika*, not just in the context of waste recycling but throughout the economy generally, will be to overcome tunnel-vision departmentalism and to create incentives for the faster incorporation of environmentally preferable technologies into the Soviet economy.

At the start of the 1990s, there existed no single agency in the Soviet Union solely responsible for managing the toxic wastes problem. It may be that the newly created agency for environmental protection, Goskompriroda, will emerge as the entity with a clear mandate to manage toxic wastes. The process for their safe management and neutralization may be clarified in the forthcoming "State Ecology Program," which was still in preparation at the start of 1990. One bit of progress is that a facility in the city of Chapayevsk (near Kuybyshev), which was originally to be used for destroying chemical weapons prior to massive public outcry, will instead be used as an education and training center for developing methods of destroying toxic substances (*Pravda*, Sept. 7, 1989, p. 1).

The environmental challenge posed by hazardous wastes is immense and growing, and few countries have adequately addressed the problem. Typically, the need to clean up hazardous situations is usually apparent only after a health-threatening incident has occurred, rather than before. The Soviet Union has no program comparable to the "Superfund" in the United States, which exists to systematically clean up hazardous sites that become identified (though it must be acknowledged that "Superfund" clean-ups in the United States have proceeded at a glacial pace). Better procedures for handling and recycling toxic wastes, and for avoiding their use in the first place, must be high on the 1990s environmental agenda of all industrialized, and industrializing, nations.

7 Managing Soviet forest resources*

The forests of the Soviet Union are vast and impressive, varying in composition from the conifers of the Siberian wilderness to the birches of the busy urban greenbelts around Moscow and Leningrad. One third of the Soviet Union is covered with trees; yet, despite the traditional value and importance of this resource to the Soviet people, their forests are disappearing at an alarming rate. Wood that is harvested is not used efficiently, and more than half the area cut down each year is not regenerated, even at a minimal level (Barr and Braden, 1988).

The Soviet Union contains about 20 percent of the world's forested area, and 25 percent of its growing stock. While the Soviet forest does not have the level of biomass of the tropical forests, it nonetheless contributes about 40 percent of the forest and woodlands of the northern hemisphere.

The Soviet Union derives many benefits from its forests. It is a major exporter of certain types of wood products, and Soviet citizens themselves consume over 300 million cu. meters of wood products each year. In addition to industrial use of wood, forests play an important role in food supply, agricultural land protection, and outdoor recreation. The environmental significance of the forest is recognized by the number and size of nature reserves and other types of preserved lands that are found within it (see chapter 8). Finally, forests have been a meaningful part of the culture, history, and folklore of many nationality groups in the Soviet Union, particularly the Russians.

The current destruction of the Soviet forest is representative of many economic and environmental issues facing not only the Soviet Union, but also most countries of the world. For example, Soviet planners must reconcile increasing consumer demand for wood pro-

* This chapter was prepared by Kathleen Braden, School of Social and Behavioral Sciences, Seattle Pacific University, Seattle, WA 98119.

ducts in an era of growing environmental awareness and understanding about the long-term cost of poor harvest practices.

Can any society achieve both economic growth and improved environmental quality? In terms of forest resources, the answer would appear to be that these apparently conflicting societal goals are reconciled through improved use of wood (more output per tree, more consumption of industrial by-products such as wood chips), consumer conservation (paper recycling), and careful regeneration of forest stands that are cut. Yet, for the Soviet government, all these measures require large outlays of capital investment in a period of increasing ruble scarcity.

In the past, logging practices in the Soviet Union appeared to be driven by the perception that "unlimited" forest reserves existed just across the Urals. Soviet forest ecologists and industrial planners now seem to agree that such beliefs are misleading, yet decisions are still made on the basis of expediency. Poor harvest and production practices continue, probably less out of misconceptions than due to the fact that investment money for new policies is simply unavailable. And the latest blow may be that just when the Soviet Union is making a better attempt to replant trees in cut areas, foresters are beginning to understand that replanting does not necessarily equal reforestation. Even a well-financed forest nursery sector, using much more advanced techniques than are now available in the Soviet Union, may be able to ensure only a future wood supply, not a recreation of the true Russian forest.

The forest resource

The 1983 inventory of forest lands in the Soviet Union showed that 810.9 million ha, or 36.4 percent of the territory of the country, is classified as "forest covered," and includes a wood volume of 85.9 billion cu. meters (*Narodnoye*, 1987, p. 249). "Forest Fund" is a broader category which encompasses about 56 percent of the national territory, taking in both forested lands and non-forested sections (land not yet regenerated with tree stock, as well as features such as roads, water bodies, and pasturelands).

Some forest covered land is not under the direct administration of central industrial agencies, but is under the control of agricultural units such as state and collective farms. Table 7.1 shows the uneven geographic distribution of 684 million ha of state forest-covered land under the jurisdiction of major forest agencies, with the bulk of the

Table 7.1 *Geographic distribution of forest covered land under central administration*

Region	Forest land (1,000 ha)	Percent of USSR total forest land	Forest as percent of total land in each Union Republic
RSFSR	657,647.4	96.1	38.5
incl. European RSFSR	12,957.2	18.9	
Ukraine	5,961.5	0.9	9.9
Belorussia	5,727.5	0.8	27.7
Kazakhstan	5,056.1	0.7	1.9
Georgia	2,146.5	0.3	30.7
Latvia	1,656.1	0.2	25.5
Lithuania	1,225.2	0.2	18.8
Uzbekistan	1,114.4	0.2	2.5
Estonia	1,043.2	0.2	23.2
Turkmenistan	937.9	0.1	1.9
Azerbaidzhan	792.8	0.1	9.1
Kirgiziia	405.7	0.1	2.0
Armenia	269	<0.1	9.0
Moldavia	243.1	<0.1	7.2
Tadzhikstan	111.2	<0.1	0.8
Total USSR	684,337.6		30.7

Source: Anuchin *et al.*, 1985, p. 522

forest (96 percent) located in the Russian republic (RSFSR). Its share is concentrated in the Asian portion of the RSFSR and in the northwest.

The geographic distribution of the forests is an important element contributing to past overcutting in some of the more accessible areas, such as the central region and along the Volga. The trend in logging has been to push the harvest into remote areas east of the Urals and into the northwestern RSFSR.

At least three major systems of forest classification are employed in the Soviet Union, based on: (1) species and age characteristics; (2) depletion level of forests; and (3) degree of protection. The first scheme is somewhat different from the silvicultural classes employed by western foresters, and is discussed below under species characteristics. The second system divides all Soviet forest regions into "forest deficit" or "forest surplus," and is a direct result of both the historic pattern of harvest or natural conditions which do not favor forest development.

Table 7.2   *Geographic distribution of group 1, 2, and 3 forests (in percent of forest fund lands)*

Region	Percent Group 1	Percent Group 2	Percent Group 3
RSFSR:			
Northwest	20.2	11.0	68.8
Central	36.1	54.6	9.3
Volga-Vyatka	20.0	45.0	35.0[a]
C. Black Earth	75.2	24.8	0.0
Transvolga	39.6	33.2	27.2[a]
Urals	19.3	38.2	42.5
Tomsk oblast	5.8	2.7	91.5
Tyumen oblast	10.0	1.8	88.2
East Siberia	12.4	0.9	86.7
Far East	11.7	3.0	85.3
Ukraine	43.2	56.8	0.0
Belorussia	28.7	71.3	0.0
Moldavia	95.0	–	–
Lithuania	41.0	59.0	0.0
Latvia	53.4	46.6	0.0
Estonia	24.0	76.0	0.0
Transcaucasus	98.6	1.4	0.0
Kazakhstan	89.8	2.8	7.4
Uzbekistan	48.2	51.8	0.0
Kirgiziia	100.0	0.0	0.0
Tadzhikstan	95.8	4.2	0.0
Turkmenia	67.7	32.3	0.0

Note: [a] In Volga-Vyatka region, only Kirov oblast has Group 3 forests
In Transvolga region, only Bashkir ASSR has Group 3 forests
Sources: Moldavia and Kazakhstan figures from: Anuchin, *et al.*, eds., vol. 2, p. 86 and vol. 1, p. 389 respectively; Tomsk, Tyumen, East Siberia, and Far East figures from: Tsymek, 1975, pp. 68, 90, 101; remaining figures from: Vorobyev, *et al.*, 1979, pp. 185, 201, 217, 230, 239, 268, 337, 345, 352, 364, 385

Forest surplus regions exist in the Urals, West Siberia, East Siberia, the Far East, and Northern European RSFSR. Forest deficit zones include: the Central region, Volga-Vyatka, Central Black Earth, Transvolga, North Caucasus, the Baltics, Transcaucasus, Moldavia, Kazakhstan, Ukraine, and Central Asia (Vorob'yev, *et al.*, 1987, pp. 119–24). Industrial harvest is supposed to be located in forest surplus regions, but in the past, over-logging in European Russia resulted in severe loss of forest stocks, and has contributed to the shift in harvest eastward and northwestward.

The third system for classifying forests is by permitted use. Group 1 is environmentally important forests protected from commercial harvest, and includes about 16 percent of national forest covered area, with a general volume of wood estimated to be 12.5 billion cu. meters. Of this volume, 5.4 billion cu. meters is considered "mature and overmature" (see age categories below). Group 1 forests make up half of the state forest area in the European Soviet Union (Table 7.2). These forests are used for urban greenbelts, resorts, erosion control, shelterbelts, road protection, and forest reserves.

Group 2 forests share both commercial and environmental protection functions, and have been overcut in the past due to proximity to industrial centers. This category makes up 58.5 million ha of forest covered area, about 6–7 percent of national total, with 6.8 billion cu. meters of wood, including 2.4 billion of mature and overmature wood.

Group 3 is the largest category, with 516.8 million ha of forest covered area, about 77 percent of the national total.* This class includes forests for commercial harvest, and has about 56.3 billion cu. meters of wood, with 42.6 billion in the mature and overmature age brackets.

Thus, in terms of conservation, Group 1 and 2 forests play the most significant role in the Soviet Union. They tend to be younger forests than those of the commercial zones, have a greater share of deciduous species, and are located more in the European regions of the country.

While precise botanical methods for classifying tree species are available, a more common method derived from industrial practices is to divide trees into hardwood and softwood species. Broadleaf trees which are deciduous (shedding leaves seasonally) are placed in the hardwood category, and needleleaf trees which are evergreen are classified in the softwood category. A notable exception is larch, which is a deciduous needleleaf tree.

Soviet foresters have a more complex system of tree classification: conifers (*khvoynye*), softleaved (*myagkolistvennye*), and hardleaved (*tverdolistvenniye*). The last two categories subdivide the broadleaf species according to whether they are shade-intolerant hardwood ("softleaved"), such as birch or shade-tolerant hardwood ("hardleaved"), such as oak. Forests of significance include: (1) conifers – pine, spruce, fir, larch, and stone pine; (2) shade-tolerant hardwoods – oak, beech, ash; and (3) shade-intolerant hardwoods – birch, aspen, linden, and alder.

---

* The 1988 State of the Soviet Environment report gives these figures as Group 1, 24 percent; Group 2, 8 percent; and Group 3, 68 percent (Doklad, 1989, p. 101).

Of the principal commercial forest area, 78 percent of the stock is conifers; 4.7 percent shade-tolerant hardwoods; and 17 percent shade-intolerant hardwoods.

Five age classes are recognized for Soviet forests: young, juvenile (literally, "polewood"), average, almost-mature, and mature/over-mature. Classifying an actual forest stand depends on variability of different age classes and species within the stand. As noted above, forests in the European USSR tend to be younger because of past exploitation, and many regions of the Soviet Union have long regrowth period requirements to bring trees up to maturity because of harsh climatic conditions. Soviet loggers consider unharvested mature and overmature trees to represent an opportunity cost to the economy; however, as noted below, such trees may play an important role in the ecological cycle of the forest.

Forests are defined as basic biotic communities, characterized by tall, woody plants. Within this biotic community, which includes associations of plants, soil, and fauna, trees compete for space and resources, until an equilibrium or "climax" vegetation type predo-minates.

The forest ecosystem is described similarly by Soviet and western sources to include:

(1) climatological factors (light, heat, moisture)
(2) soil and forest litter
(3) biotic factors (animals, plants, microorganisms)
(4) anthropogenic factors (modifications by human beings)

As trees compete in the forest ecosystem, one type eventually domi-nates to produce the climax forest, but forests may often be quite mixed; that is, in different stages of the process of *invasion and succession* of species.

Soviet foresters recognize two types of this invasion-succession pattern. The first, self-generative, requires a long time to complete a cycle. Shade-intolerant species, such as birch, pine, and aspen may come into a region first, followed in time by shade-tolerant trees such as spruce, fir, or linden. The second type of succession is termed "exogenously generated" because it is a result of climate changes, catastrophic events, or interference by human actions. For example, in the large northern coniferous forest referred to as *taiga*, logged regions which were forested with native spruce and pine may be succeeded by birch, aspen, or alder (Figure 7.1). Since most harvested forest lands

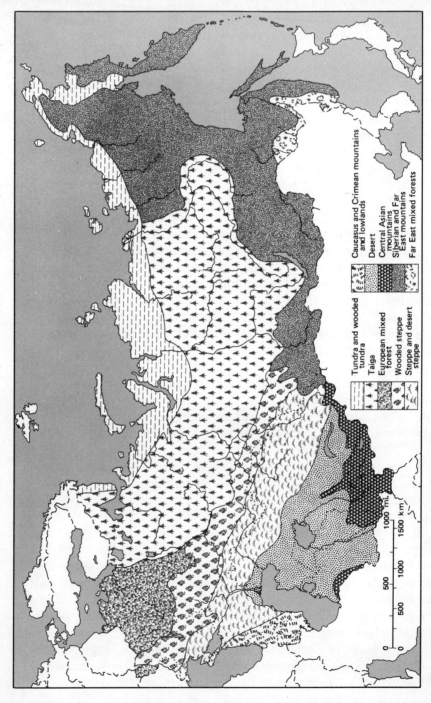

Figure 7.1   Natural zones of the USSR *Source:* Adapted from Mil'kov, 1977

Tundra and wooded
tundra

Taiga

European mixed
forest

Wooded steppe

Steppe and desert
steppe

Caucasus and Crimean mountains
and lowlands

Desert

Central Asian
mountains

Siberian and Far
East mountains

Far East mixed forests

are left to natural regeneration, deciduous species seem to be replacing much of the nation's softwood resources.

The Soviet forests provide important habitat for smaller plants and many animals, with more than 100 species of mammals, 300 species of birds, 40,000 insect species, and 15 reptile species. These animals not only contribute to enjoyment of outdoor recreation; they also are part of the forest ecosystem. For example, cone eaters, such as the brown bear, may help redistribute seedlings and propagate the forest. Animals enrich forest soils with dung, control plants, and serve as a vital link in the natural chain of forest life.

Soviet system of forest management

Forest industries in the Soviet Union have treated the forest resource as a type of "mining" operation, in which trees are cut down and then the logging brigades move on to new stands of trees, ever further away from central regions. The idea of sustained yield forest harvest, in which the amount cut corresponds to the annual growth per year, has not yet been practiced in the Soviet Union, with a few recent exceptions, such as some enterprises in the Ukraine or Baltic republics.

Several major agencies manage the forests of the Soviet Union. Minlesprom, the Ministry of Forest Products Industries, newly re-created in 1988 by consolidating the operations of the old Ministry of Pulp, Paper, and Woodworking, has traditionally overseen commercial use of wood. Silvicultural activities were also reorganized in 1988 and are now under the management of the State Forestry Committee, Goskomles. However, the 1988 reorganization apparently gave a greater share of management responsibilities to the forestry department of Minlesprom and the newly-created State Committee on Environmental Protection (Goskompriroda).

The new management scheme is an attempt to tie forest management more intimately into forest utilization. The revised structure envisions *perestroika*-style regulations of economic accountability and profit-making.

Management of forests in the Soviet Union has historically been a controversial issue, and the industry's ministry has been reorganized many times in an attempt to improve efficiency and hold upper-level managers responsible for failure to achieve goals. As a recent example, the Minister of Minlesprom, Mikhail Busygin, was dismissed from his post in 1989 for failing to achieve environmental improvement and production goals. A basic aspect of the problem is the fact that the forest ministry does not receive a large share of investment (about 4 percent of national basic fund investment in all industry in 1987), and Gosleskhoz has been allocated even less capital than the wood product enterprises (Petrov *et al.*, 1986, p. 211). Without substantial new investment, industry and environmental targets may be difficult to attain.

The Russian term, *lesnoye khozyaystvo*, meaning "forest management" is used as an umbrella expression to include all aspects of governing forests, from silviculture to forest fire prevention. The phrase is taken to be distinct from *lesnaya promyshlennost'*, forest industry, which is concerned with large, centralized, commercial uses of the forest. Under Soviet laws, the forest management branches have the following responsibilities:

(1)   management of forests to strengthen water conservation, regulate microclimate, improve the human environment
(2)   provision of trees for industrial harvests
(3)   increase and improvement of species composition and quality of trees and their products
(4)   protection of forests from fires, illnesses, and other harmful factors
(5)   to use State Forest Fund lands rationally
(6)   improvement of technology and science of forest management (Tunytsya, 1987, p. 91)

Despite this administrative structure, the amount of forest land under any kind of real management in the Soviet Union is quite small. For lands in the total forest fund, the broadest category of management attention is termed *lesoustroistvo* ("forest regulation") and may include a range of activities, from direct conservation measures to aerial and ground surveys of forest stands for mensuration and classification. Actual intervention for timber stock management is termed *lesnye kul'tury*, or "forest culture," and nationally is only 2 percent of forest covered lands. On the other hand, 45 percent of forest covered lands in the Ukraine are receiving forest culture management (Anuchin, 1985, p. 522).

Forest culture should not be interpreted to mean tree replanting, however, because "natural" methods of forest regeneration may be included (see Table 7.6 below, p. 131). In 1987, 2 million ha nationally were in a category called "forest restoration" (*lesovosstanovleniye*); but of this amount, only 987,000 was seeded or plantation restoration. The remainder, "assisted natural regeneration" refers most often to strips of trees left after a plot is logged.

From 3.5 to 4 million ha of forest per year are cut in the Soviet Union; therefore the amount of aggressive forest restoration is only 25 to 28 percent of the amount harvested.

Using the Soviet forests

The forest products industries of the Soviet Union do not make a contribution to the national economy commensurate with the size of the forest resource, nor do they as a whole contribute a large share to the export base of the Soviet Union, except in selective cases, such as log exports to Japan. Forest industries make up about 5 percent of gross industrial output, and 7 percent of the value of exports from the Soviet Union (Barr and Braden, 1988, pp. 102 and 151). The amount of industrial product derived from each tree is significantly lower in the Soviet Union than is the case in Scandinavia and North America. Much Soviet wood is wasted in the production process, and a large share of national harvest still goes for fuelwood (20 to 22 percent of the yearly cut). Improvements in the industrial use of wood could be an important first step toward better conservation of the Soviet forest.

Shortcomings in forest products industries tend to have a negative impact on the environment of the Soviet Union in four ways.

(1) Poor use of deciduous trees in industrial processes. In the pulp and paper industry, for example, deciduous species make up only 9 percent of the raw material base, versus 30 percent in the United States and 58 percent in Japan (Tunytsya, 1987, p. 185). Improved management and consumption of deciduous species under a sustained yield forestry system in the European regions of the Soviet Union could reduce the need for shifting timber harvesting operations to Siberian stands or overcutting accessible coniferous forests.

(2) Poor use of wood by-products. More than 50 million cu. meters of leftover wood material is produced yearly in the Soviet Union during industrial processes, but only half of it is used to make composition boards or contribute to the wood chemicals sector. Twenty percent of all wood harvested in the Soviet Union is used directly to make packaging material, such as paperboard, whereas much of the material needed could come from the woodchip by-products of sawmilling and other processes (Tunytsya, 1987, p. 58).

(3) Forest industries as direct sources of environmental damage. Deforestation has contributed to soil erosion and water resources have suffered due to effluent discharges from pulp and paper mills, with only 19 percent of the pulp and paper industry's waste water receiving effective treatment. Water shipment of logs is giving way to rail and truck transportation in the Soviet Union, but in 1980, 53 million tons of timber still moved by water, and submerged logs are very harmful to river ecosystems, and to transportation.

(4) Low level of capital investment in forest industries. One approach to improving both the environmental record and output performance of the forest sector is in creating new and cleaner technologies. Such processes would make better use of trees, allow selective cutting, enhance replanting efforts, raise productivity in industry, and reduce pollution levels. Unfortunately, such technological shifts come at a high price, and investment levels in forest products as a whole have not grown since the 1950s. In 1970, for example, forest industries were allocated a 5.1 percent share of national basic fund investment in industry; in 1980 this share was 4.5 percent, and in 1986, 4.2 percent. In fact, forest products activities, especially logging and sawmilling, are probably some of the most severely under-capitalized of any Soviet industry.

The Soviet forest products industry thus faces the formidable three-fold task of improving production efficiency to satisfy consumer demand, offering more Soviet wood products on world markets to provide export income, and improving wood utilization practices which have already had severe consequences for the Soviet environment. All this is to be accomplished in a backwater part of a troubled economy.

Industry is not the only source of demand on the Soviet woodlands. Recreational forests exist in Group 1, 2, and 3 forest zones, and are specially designated for the enjoyment of the population. Approximately 32 million ha of Group 1 forests are listed as of direct recreational significance. In addition, tourist and hiking forest belts are present throughout some Group 2 and 3 forests zones; and some categories of reserved forest lands, such as national parks and hunting reserves, provide recreational forests.

Health resorts are common in forest areas. There is an uneven geographic distribution of forest lands devoted to health resorts, with the Caucasus region playing a particularly important role (56 percent of the total).

Forests may have a combination resort and education function. For example, in Belorussia, half a million people annually are taken on forest tours for environmental education, under the central control of the Botanical Garden of the Belorussian Academy of Sciences (Bakhar', 1984, p. 71). In addition, 6,000 hunting clubs exist in the Soviet Union, with access to 291 million ha of forest for sport hunting and fishing. Commercial hunting enterprises, as well as 420 special Forest-Hunting Management units, also make use of forest lands.

Many resort or recreation forests are connected with greenbelt zones around urban areas, totalling 12 million ha in the Soviet Union. In and around Moscow, for example, 172,000 ha of protected forest zones are available for recreation, including one national park (see chapter 9). This produces an intensive ratio of population to forest land (110 people per ha of recreational forest park). Greenbelts are part of the Group 1 forests, and also are designed to contribute to the environmental health of urban residents through microclimate amelioration and water conservation. Seventy-five percent of greenbelt forests are located in the European regions of the Soviet Union, reflecting the urban orientation of this forest type. Examples of the extent of urban greenbelts in Soviet cities are shown in Table 7.3.

Forests have always served as a source of food for people of the Soviet Union, often tied in to recreational functions, such as berry picking. More than 5 million tons of edible products per year are derived from the forests. Pasture areas, cattle grazing regions, and apiaries for honey production are also part of the Soviet forest lands, and wild nuts and fruits are a valuable forest commodity, especially in Central Asia and the Caucasus. Walnuts, pistachios, almonds, chestnuts, beechnuts, hornbeam, apples, pears, apricots, and olives are among the most common tree products. The medicinal value of plants from the forests is held in high regard in the Soviet Union, and over 600 types of forest plants are harvested for the pharmaceutical industry. Even certain mosses from the forests have a use for Soviet society: Crimean oak trees yield a moss used for aromatic extraction to make a brand of perfume called "Forest."

Tree plantings (often termed "shelterbelts") are used in the Soviet Union to protect or buffer roads, soils, agricultural activities, and water bodies. More than 3 million ha of forest belts are planted alongside roads in the country, and are used to absorb emissions from automobiles (Figure 7.2). Forests are conserved along the shores of 1,500 rivers and lakes in the Soviet Union as protective vegetation belts. The Soviets have even designated within Group 1 forests a

Table 7.3   *Greenbelt areas in selected Soviet cities (data as of January 1, 1989)*

City	Hectares	% of city	Population	ha per 1,000 people
Alma-Ata	4,637	17	1,128,000	4.11
Baku	9,628	44	1,150,000	8.37
Dnepropetrovsk	15,376	39	1,179,000	13.04
Donetsk	18,223	51	1,110,000	16.42
Frunze	4,946	39	616,000	8.03
Gorkiy	9,532	29	1,438,000	6.63
Irkutsk	11,514	38	626,000	18.41
Kiev	57,630	70	2,587,000	22.28
Kishinev	3,444	21	665,000	5.18
Krasnodar	3,776	22	620,000	6.09
Krasnoyarsk	8,615	25	912,000	9.45
Krivoy Rog	13,656	33	713,000	19.15
Kuybyshev	16,472	35	1,257,000	13.10
Leningrad	15,318	27	4,456,000	3.44
Minsk	5,131	25	1,589,000	3.23
Moscow	21,509	22	8,769,000	2.45
Novosibirsk	20,706	43	1,436,000	14.42
Omsk	10,878	25	1,148,000	9.48
Perm	35,865	50	1,091,000	32.87
Riga	6,632	22	915,000	7.25
Saratov	6,552	17	905,000	7.24
Tallinn	3,900	25	482,000	8.09
Tashkent	10,011	37	2,073,000	4.83
Tbilisi	9,983	29	1,260,000	7.92
Tolyatti	9,879	33	630,000	15.68
Ufa	23,984	51	1,083,000	22.15
Vilnyus	15,399	54	582,000	26.46
Volgograd	12,133	28	999,000	12.15
Yaroslavl	3,037	17	633,000	4.80
Yerevan	5,059	24	1,199,000	4.22
Zaporozhe	13,606	44	884,000	15.39

Source: *Okhrana*, 1989, pp. 18–19; last column calculated by author

45 million ha line of protective trees along the southern borders of the tundra to reduce the impacts of severe weather on regions to the south.

Some of the protective forest belts are sections of true forests; in other cases, they are less "forests" than lines of trees designed to cut wind or anchor soil moisture. All such belts, however, are considered to be an important part of the state forest fund in the Soviet Union.

Soviet scientists recognize the role forests play in the oxygen cycle and other environmental processes. Since the Soviet forests represent a large portion of the biomass in the northern hemisphere, continued

Figure 7.2 Typical shelterbelt protecting wheatfields near Kursk

depletion of the resource may have harmful effects on the planetary oxygen supply. The ability of trees to absorb harmful air pollutants is also well-respected in the Soviet Union: one hectare of 20-year-old pine forest is believed to absorb daily 9.35 tons of carbonic gases and give out 7.25 tons of oxygen (Tunytsya, 1987, p. 57). Aggressive tree planting programs in cities have taken place to use these pollution-absorption capabilities of trees. Soviet scientists believe that forests mitigate the impacts of noise pollution. For example, each 100 m of forest (especially deciduous trees) reduces noise by 20 decibels. Urban greenbelts and tree zones along roads, therefore, also serve the function of making traffic and other noise less harmful.

In short, forests provide multiple benefits to Soviet society, far beyond their use in wood, paper, and chemical industries alone. Unfortunately, even though the Soviet Union recognizes the value of preserving its tree resources, the ruin of forest lands continues.

Causes of forest destruction

Forests are disappearing from the Soviet Union, but the phenomenon is not evenly distributed in terms of geography. Some regions, particularly in the European areas and accessible districts of the Far

East, have been damaged much more than others. While natural factors account for some of the destruction, the prime cause is poor management of human activity.

The very nature of the forest ecosystem connotes that a certain volume of mature and overmature trees dies each year in the forest. Within the Soviet Union, 51 million cu. meters of wood is in the "mature" age category. When the oldest portion of this wood is not harvested, it may be regarded as lost to the state economy. On the other hand, the death of these trees does make a contribution to the natural cycle of the forest, and thus may be seen as contributing an environmental benefit.

Forest fires, on the other hand, can exacerbate deforestation, although many naturally occurring fires play a positive role in forest regeneration. The problem in the Soviet Union is the number of fires caused by human beings (from 70 to 85 percent – up from 44 percent in 1966), and the large scale of some fires, particularly in Siberia where dry and inaccessible coniferous forests are vulnerable in the summer months. From May to September, 1915, for example, a forest fire covered an area of 1.6 million sq. km, from Tobolsk to the Lena River. More recently, up to 15 percent of the forested area of the Soviet Far East was reported lost to fires in the 1960s. Other large fires in the taiga took place in 1925, 1928, 1947, 1962, and 1987, and in the spring of 1989 vast fires burned 20 percent of the forests – over 200,000 ha – on Sakhalin Island (*Pravda*, June 30, 1989, p. 2). For all of 1988, 24,400 fires burned a total of 792,400 ha of Soviet forests (*Okhrana*, 1989, p. 114).

Of animal species that are possibly harmful to the forest, the most bothersome is considered to be insect pests. For example, red-headed fir leafroller caterpillars have been particularly damaging to Siberian fir forests, with an outbreak in 1970 spoiling 80 percent of the fir cones in Irkutsk oblast. Siberian moths, the large black fir beetle, and the fir bark beetle also cause severe forest loss. From 1942 to 1946, and again in 1952, a massive Siberian moth infestation occurred in various parts of the Soviet Far East.

The Soviet Union does not publish details on acid rain damage to the national forest cover. Information may be estimated, however, from foreign sources, Soviet data on air pollutants, and comparable experience in other northern hemisphere countries.

The sources of acid rain (or "acid precipitation"), such as burning fossil fuels and driving motor vehicles, have been discussed in chapter 2. Portions of forests are damaged by this acidic precipitation, as well as by accompanying acidification of soils. Needles or leaves become yellow and fall off, the tops of trees thin out, trunks are deformed, and even roots may be harmed.

The Soviet Union, with 25 million tons per year of sulfur dioxide emissions, is ranked as the number one producer in the world (Dovland, 1987). Soviet data indicates that the Soviet Union produced 18.6 million tons of sulfur anhydrides in 1987, 4.5 million tons of nitrogen oxides, and 15.5 million tons of carbon monoxide (*Narodnoye*, 1988, pp. 571ff).

Output of these pollutants alone, however, does not determine acid rain damage to vegetation and water bodies. The geographic pattern of pollution dispersal and the species of forest affected are also important factors. Of the amount of acid rain estimated to be deposited on the Soviet Union, 32 percent is believed to originate outside the country, probably in the industrial states of northern and eastern Europe, and the remaining 68 percent is believed to be produced within the Soviet Union. Many large industrial cities within or near forest zones contribute large amounts of acid rain causing pollutants (see Table 2.3 and Appendix 2.1). For example, the smelters and other sources of emissions around Norilsk have degraded 5,450 sq. km of forests (Doklad, 1989, p. 149).

The Soviet Union ratified the Convention on Long Range Transportation of Air Pollution in May, 1980, and in 1988 promised to achieve a 30 percent cutback in sulfur dioxide emissions by 1993. Such a reduction may be important, because in European and North American areas with similarly high levels of emissions, damage to forests, particularly pine, has been quite severe. It is not unreasonable to assume that vast tracts of Soviet northern forests are being destroyed likewise by the acid rain phenomenon.

Pollution is not the only source of human damage to the forests. The Soviet population is increasing its use of forests for outdoor recreation, with growing pressure on urban greenbelts. Visitors tend to "love the forests to death" when large numbers of people descend on outdoor areas that have inadequate controls on camping and hiking activities. The resulting problems can range from trampling down the vegetation and illegally removing trees for firewood, to poor tending of campfires and a subsequent increase in forest fires. In an effort to establish guidelines for recreational forest use, a meeting was held in May, 1985, sponsored by the State Forestry Committee. The main issue was how to reconcile conservation and recreational use of the forest. Soviet planners recognized that norms need to be established for the carrying capacity of heavily used forests near

Table 7.4   *Wood harvest, 1970–1987, selected years*
(mill. cu. meters)

	1970	1975	1980	1985	1987
Total harvest	385.0	395.1	356.6	367.9	389.2
Fuelwood	22.5%	21%	22.2%	23.6%	22.2%

Sources: Narodnoye khozyaystvo SSSR v 1987 godu and *Nardonoye khozyaystvo SSSR v 1978 godu*, pp. 142–43 and p. 165 respectively

urban areas, depending on the season, type of use, and type of forest (Bakhar', 1984, p. 67).

Pollution and irresponsible recreational use create much harm, but the single most damaging activity measurable today is probably logging. Since 1917, a shortfall of approximately 138 million ha of unreplaced forests has been created in the Soviet Union (Barr and Braden, 1988, chapter 3). One of the prime causes of deforestation has been violation of allowable cut norms, but other reasons include poor logging practices, insufficient use of deciduous species, insufficient use of thinning cuts, and lack of efforts to regenerate forests.

Soviet foresters do calculate norms of allowable cut per year, and the Soviet Union is probably capable of cutting a maximum of 400 million cu. meters per year (compare Table 7.4). The distribution of harvests, however, suggests that allowable cut norms have been violated severely in some regions (Figure 7.3). Overcutting of coniferous stands of trees in accessible zones such as the Baltics, Belorussia, Central Region, the Ukraine, and the Transcaucasus occurred through the 1960s. Some regions suffered from overharvesting through World War II, and the Urals were overcut in the 1950s.

The commercial harvest of wood for selected years from 1970 to 1987 is shown in Table 7.4. The published figures for wood harvest, however, underestimate the actual amount of forest cut because they do not show a category termed "intermediate harvest," usually wood from the thinning of forests or from collective farms. Because this category is not reported as part of commercial harvests, total volume of wood cut may be underestimated; for example, Barr and Braden (1988, p. 58) conclude that the 1984 harvest was larger by 11 percent than the figure indicated by industrial cut alone. In 1987, 57 million cu. meters were listed as intermediate harvest, or 14.6 percent the amount of commercial removals (*Narodnoye*, 1987, p. 250). This "extra" harvest

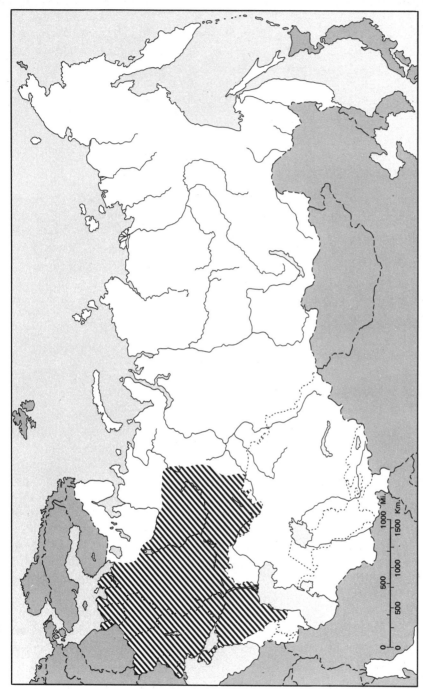

Figure 7.3 Area of past forest overcutting *Source*: Barr and Braden, 1988, map provided by K. Braden

can be regionally significant (for example, half of wood removed in Latvia and the Ukraine is intermediate cut), and is derived from all three groups of forests because it is considered a silvicultural measure to enhance forest growth.

Total harvest of trees has shifted regionally in the Soviet Union as forest stocks in the European regions have become depleted. The last published figures on harvest by economic region were published in 1975, and showed that 34.3 percent of total RSFSR industrial logging took place in West Siberia, East Siberia, and the Far East, while those regions accounted for 24.7 percent of the harvest in 1960. This geographic shift to less accessible areas has resulted in long-distance hauls of raw materials to industrial processing facilities closer to the source of consumer demand. In addition to raising the cost of logging, decreasing the profitability of the sector, and increasing energy and pollution loads, the shift has decreased the incentive to make more efficient use of forests remaining in the European regions.

Harvest practices in the Soviet Union have not traditionally been kind to the forest. Clearcutting (removing all trees in the cutting area) is common, and many types of undesirable species are left in the cutting area or burned on site. Much logging is done by small brigades of workers in remote sites, with little reward for environmentally sound logging methods (Barr and Braden, 1988, chapter 3).

Forest restoration

Saving the forests of the Soviet Union is not just a matter of replanting trees, but it would be a good start. As noted above, the amount of trees replanted averages about 25 percent of the harvest. Furthermore, some afforestation figures must be interpreted conservatively because they reflect new shelterbelt plantings of trees to enhance soil protection in regions such as Turkmenia.

Total afforestation (creating and recreating of forests) averages about 2 million ha per year, with 85 to 95 percent located in the RSFSR (Table 7.5). Regeneration of forests would therefore appear to correlate highly with commercial logging. However, when afforestation is broken down to two types, replanting and naturally assisted regeneration, a different picture emerges. The geographic distribution of replanting does not necessarily correspond to areas of most intense logging. As demonstrated in Table 7.6, of the amount regenerated nationally in 1987, less than half was accomplished through replanting or reseeding. While the RSFSR accounted for 96 percent of national

Table 7.5   *Soviet afforestation, by region (selected years, 1,000 ha)*

Area	1970	1980	1985	1986	1987
USSR incl.:	1,997.3	2,178.7	2,188.5	2,189.5	2,202.7
RSFSR	1,731.4	1,862.3	1,874.5	1,879.5	1,893.9
Ukraine	59.9	46	44	39.7	38.1
Belorussia	30.8	33.8	28.1	29.1	29.3
Uzbekistan	31.8	29.4	46	46.2	46.1
Kazakhstan	50	91.6	83.8	83.1	83.4
Georgia	20.9	28.1	29.4	29.4	29.6
Turkmenia	24.4	23.7	35.9	35.8	38.3
Estonia	12.3	14.7	9.4	9	8

Source: Narodnoye . . ., various years

Table 7.6   *Type of forest regeneration 1987 (selected regions, 1,000 ha)*

Region	Total	incl. planted	Region share of total planted, percent	incl. natural	Region share of total natural, percent	Percent planted	Percent natural
USSR incl:	2,202.7	987.3		1,215.4		44.8	55.2
RSFSR	1,893.9	726.5	73.6	1,167.4	96.1	38.4	61.6
Ukraine	38.1	35.8	3.6	2.3	0.2	94.0	6.0
Belorussia	29.3	26.1	2.6	3.2	0.3	89.1	10.9
Uzbekistan	46.1	45.1	4.6	1	0.1	97.8	2.2
Kazakhstan	83.4	73.3	7.4	10.1	0.8	87.9	12.1
Georgia	29.6	5.5	0.6	24.1	2.0	18.6	81.4
Turkmenia	38.3	38.3	3.9	0	0.0	100.0	0.0
Estonia	8	7.6	0.8	0.4	.0	95.0	5.0

Note: "percent planted" and "percent natural" indicate: of total regeneration in region, percent in each activity
Source: Narodnoye, . . ., 1988, p. 249

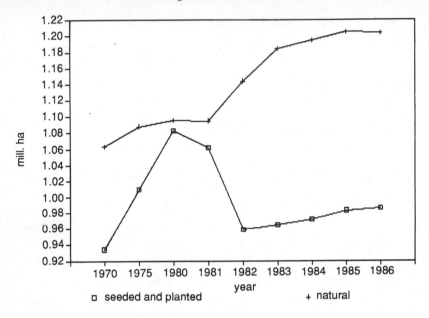

Figure 7.4   Afforestation area, USSR

afforestation efforts, it only showed a 74 percent share of national replanting work. The Ukraine, bearing the scars of past overcutting, and with a smaller forest area and share of commercial logging than the RSFSR, was much more aggressive in terms of replanting and reseeding trees (Figures 7.4 and 7.5). In fact, the Ukraine could now serve as a model of improved attempts to regenerate forests in the Soviet Union. Enterprises in this republic, particularly in the Carpathian regions, have experimented with nurseries, reseeding, and the introduction of new species, such as Douglas fir.

Volynskiy oblast in particular, with a commercial forest area of 420,000 ha, has been the focus of nursery and greenhouse activities to restore the overcut forest. Enterprises in this region, as well as other parts of the western Ukraine, are combining sustained yield logging, plantation forestry, and broader definitions of the potential harvest from forest stands to include food as well as fiber. The forest complex in Volynskiy oblast derives more than 1 million rubles worth of secondary products from the forest each year, including mushrooms, berries, and other food or medicinal materials. Thus, one complex, the "Prikarpatles" union, averaged a profitability level (ratio of unit price to unit cost) of 22 percent from 1982 to 1985, more than twice the national level for forest products activities in general, and much higher

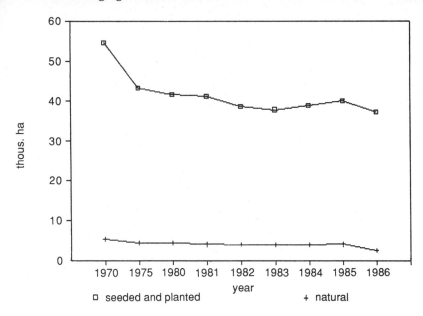

Figure 7.5   Afforestation area, Ukraine

than most logging enterprises. However, the level of investment in Prikarpatles was also much higher than the national trend in forestry, an average 8.5 percent per year at the national level in forest products (Tunytsya, 1987, pp. 63ff). Experience in the Ukraine suggests that when enough rubles are available and combined with management and technical expertise, regeneration of tree stands leads to profitable and rational use of forests.

Replanting needs to be accomplished on a much larger scale in the Soviet Union, but the result may be merely new tree farms, not renewed forest ecosystems. A study conducted by a team from the United States Forest Service has submitted that old growth forests may have certain characteristics which are not replicated by the type of plantation forest that is profitable for industry. For example, old growth (or virgin) forests have a multilayered canopy, rather than the even canopy that tends to occur with planted stands. This multilayering effect creates different light paths into the forest, and therefore supports a separate type of undergrowth component to the forest than would a replanted area. Also, in old growth forests, trees represent a range of ages, and dead trees provide nutrients to start the food chain, as well as "nurse logs" for vegetation regrowth. In short, replanting trees does not necessarily lead to regenerating a forest ecosystem, and the term "reforestation" may be a misnomer.

If this research is correct, what are the implications for preserving forests in the Soviet Union? At present, the backlog of forest lands waiting for any replanting at all is so large that it may already be too late for true forest preservation in harvested regions. The best that could be accomplished in the European Soviet Union and parts of the Far East may be replanting of stands to lead to a sustained yield "cropping" of trees, combined with preservation of some greenbelts for controlled recreational use. Conservation of whole forest ecosystems may be best planned for regions only recently tapped for logging: Siberia and the northwest. Such a scheme, however, would go exactly against current trends in the geographic shift of tree harvesting.

Choices to save the Soviet forests

Despite all the problems described in this chapter, there is some basis for optimism about the future of the Soviet forests. First, a growing environmental awareness among the people of the Soviet Union has led to more demands for forest preservation and increased pressures on industrial ministries to take better care of forest resources. Second, technical literature appearing in the Soviet Union demonstrates a new level of sophistication among forestry experts that the old-style approach to exploiting the forests can no longer be supported on either an ecological or economic basis. Third, a system for conserving forest lands is already in existence and can be strengthened (Group 1 forest classes, *zapovedniki* and other types of preserved lands; see chapters 8, 9, and 11). Fourth, *perestroika* may mean more incorporation of advanced silvicultural techniques from abroad, a greater exchange of information, and management training for Soviet foresters, as well as possible future joint ventures to care for and harvest the Soviet forests. Many challenges, however, will have to be met by the Soviet Union to ensure that the past destruction of its forests does not continue into the future. Table 7.7 summarizes what needs to be done, and the possible complications which will arise in accomplishing the tasks.

Finally, conservation of the remaining Soviet forest may come down to an issue of economics: what are the Soviet people willing to pay for the preservation of pristine forest ecosystems, often far from major population centers and inaccessible for even direct non-consumptive

Table 7.7 *Management challenges to preserve the Soviet forest*

Task	Difficulty
More efficient use of every tree harvested	Requires scarce investment capital to modernize logging sector
More use of industrial by-products (wood chips)	Will require large scale investment to "retool" forest products industries
Better use of deciduous species, notably birch, and of smaller diameter trees	Requires introduction of new technologies in some cases
Better management of recreational use of forests	Increasing demand for outdoor recreation and on-site environmental education experiences
More value-added for forest exports	Immediate need for hard currency; competition from domestic demand
Reduction of acid rain impacts	Investment cost of energy substitutions, increasing opposition to nuclear power, cost of anti-pollution equipment
Increased replanting and reseeding of forest stands	Cost and technology requirements, awareness that replanting does not necessarily equal reforestation

use, such as recreation? Forest experts in the Soviet Union are already discussing the question of economic valuation of ecological functions of the forest, and the issues confronted resemble similar problems in the market economy of the west. When the value of a log delivered at the mill is so clearly accounted, even in a system of subsidies and centrally derived prices, how can society measure the value of the uncut tree that absorbs air pollutants, provides a pleasant camping environment, or serves an important function in an ecological cycle?

These are universal issues, but the Soviet Union is facing such questions in a time of extraordinary economic and social change. In moving toward a more market-style economy, Soviet forest planners may find that, while the traditional Soviet model has failed, other systems have fared little better in the search for wise use of wood resources. If it is not too late, the Soviet Union may therefore be pressed to create an entirely new management paradigm for the forest. Such a breakthrough would have significance far beyond the borders of the Russian taiga.

8 Soviet nature reserves (*zapovedniki*)

Stretching from the Baltic shore to the sands of Central Asia to the ice of Wrangel Island, the USSR's network of state nature reserves, or *zapovedniki*, represents that country's principal biogeographic preservation effort. The *zapovedniki* system was initiated shortly after the revolution, and by 1990 encompassed around 160 reserves with a combined area of over 22 million ha. They can be found in all the major natural zones of the country, and the individual reserves range in size from 117 ha to well over a million.

The nature of Soviet *zapovedniki*

The primary purpose for which *zapovedniki* are created is to serve as scientific research stations and to engage in biosphere conservation activities. Individual reserves, however, differ in the nature of their primary scientific and conservation functions.

Several of the older *zapovedniki*, the majority of them located in the European part of the country, are intensively used for scientific research and publish their own volumes of collected works. On the other hand, some of the more recently established reserves of great size in remote portions of Siberia and the Far East may as yet have few permanent on-site research facilities, and function mainly as centers of biosphere protection or *de facto* wilderness areas. An important research function in those with permanent research stations is general biosphere monitoring, especially if nearby economic activities have the potential to adversely effect flora, fauna, or soils. It is not uncommon for some of the research done in the reserves to have direct economic applications, such as improving soil fertility, controlling pests, introducing or improving commercial animal species, etc.

Many of the reserves have been established to preserve representative samples (sometimes called "standards of nature") of a specific type of landscape or natural zone, such as steppe, estuary or montane ecosystems. To this end, several of the *zapovedniki* are made up of a number of discontiguous sections, sometimes separated by great distances, in order to preserve important outlying exclaves of biotic importance. This is particularly the case with several reserves in the European steppe biome.

Related to this, a common theme in a number of the reserves is the effort to preserve or re-establish rare or endangered species of flora or fauna. In the vast majority of them, a primary objective is to protect the breeding or wintering grounds of wildlife, primarily mammals and birds, and to protect key migration routes for the latter. A few *zapovedniki* have geological features (caverns, hot springs, minerals, etc.) as their primary preserved feature. The goal of habitat protection is a strong secondary objective in all of those reserves where it is not the primary function.

The nature reserves are not viewed as having any significant tourist function, and indeed the term *zapovedniki* itself derives from a word meaning "restricted" or "forbidden." However, many of the reserves have museums or displays that are intended to serve a local conservation education role (Aralova and Zykov, 1984). A very small number of the older ones, such as Stolby near the Siberian city of Krasnoyarsk, have traditionally provided a large-scale recreation opportunity, and are thus exceptions to the general rule. Unlike national parks in the United States, no effort has been made to include all the most remarkable natural features of the country in the system, it apparently being assumed that the nationalization of the land precludes any need to do this.

New *zapovedniki* are created based on recommendations by scientific bodies or governmental agencies within the fifteen Union republics, with the concurrence of the appropriate planning bodies. They are managed by a variety of scientific and governmental organs, including the Academy of Sciences, the Central Directorate for Conservation, Nature Reserves, Forestry and Hunting under the USSR Ministry of Agriculture, the RSFSR Central Directorate for Hunting and Nature Reserves, republic-level committees for environmental conservation or forestry, and several republic-level ministries for forestry or agriculture. In the late 1980s, responsibility for the management of most of the *zapovedniki* was reportedly transferred to the new State Committee on Environmental Protection (Goskompriroda), and all of the reserves in Uzbekistan have been put under the Uzbek republic Goskompriroda.

At present, no national agency exists to give unified management direction to the entire assemblage of reserves. There may be an intent for Goskompriroda to play this role; this question will be examined later in this chapter. The scientific work of the *zapovedniki* is coordinated to a certain extent by the All-Union Institute of Nature Conservation and Reserves which was established in 1979 within the USSR Ministry of Agriculture (Gavva *et al.*, 1983).

History of the *zapovedniki* system

The statute creating the *zapovedniki* system was signed by Lenin on September 16, 1921 (for its text, see Pryde, 1972, p. 213). Prior to the revolution several large *de facto* preserves existed in the form of hunting estates maintained by the nobility. Some of these, such as Belovezhskaya Pushcha and Kavkaz, were redesignated as *zapovedniki* after 1917. In addition, just prior to the revolution, a few nature reserves were created whose purpose was very similar to that described above for contemporary *zapovedniki* (e.g., Suputinka in 1911, Lagodekhi and Morittsala in 1912, Barguzin and Kedrovaya Pad in 1916). Starting with Astrakhan in 1919, new reserves were steadily created throughout the Soviet Union's first three decades. However, under Stalin, the purposes of the *zapovedniki* were changed from "inviolable" areas intended to study and protect nature, to areas on which scientists would learn to master and transform nature to serve the needs of the economy (Weiner, 1988). The problem was especially acute at the Askaniya-Nova reserve in the Ukraine (Weiner, 1988, pp. 70–78; Boreyko, 1990). Nevertheless, by the early 1950s, an impressive total of 128 *zapovedniki* existed (Table 8.1).

In 1951, with the anti-preservation philosophy having become well entrenched, a major reduction in the *zapovedniki* system took place. The government, responding to perceived economic priorities, and undoubtedly with the approval of Stalin, reduced the number of reserves by almost 70 percent and their area by 88 percent. Although most of these reserves have since been re-established, the continuity of the protected regime, and of the scientific activities carried out on them was lost (Figure 8.1). In addition to the many reserves that were abolished, numerous others that were not eliminated, such as Kavkaz and Sikhote-Alin, had their size (and biological integrity) significantly reduced. The total area in reserves that had existed in 1950, over 12 million ha, was not to be achieved again until the mid-1980s.

A companion category of semi-preserved areas, called hunting preserves (*zapovedno-okhotnich'ye khozyaistvo*) was started in 1957,

Table 8.1 *Growth of the zapovedniki system*

Year or date		Number	Area (ha)
	1925	9	984,000
	1933	21	4,085,916
Apr.	1937	37	7,138,300
	1947	91	c. 12,000,000
	1951	128	c. 12,500,000
	1952	40	1,465,688
Sep. 1	1959	76	3,296,000
Jan. 1	1961	93	6,360,000
Mar. 1	1964	66	4,267,400
Jan.	1969	86	6,700,000
May	1976	107	8,960,000
Jan. 1	1981	128	10,783,127
Jan.	1988	155	18,886,223
(est.)	1990	160	22,500,000

Sources: 1925 and 1933: Weiner, 1988, p. 243; 1937 through 1964: Pryde, 1972, Table 4.1; 1969: Bannikov, 1969; 1976: *Zapovedniki . . .*, 1976; 1981: Borodin and Syroyechkovskiy, 1983; 1988: Drucker and Karpowicz, 1989 and Appendix 9.1; 1990: author's estimates.
Note: The 1989 environmental statistical handbook indicates a total area in *zapovedniki* on Jan. 1, 1989 of 21,374,000 ha. Among the new units are a large Putorana reserve in East Siberia of 1.89 million ha. Three other new *zapovedniki* are shown for the Russian Republic, as well as one new unit in each of the Armenian, Azerbaidzhan, Moldavian, Turkmen, and Uzbek republics (*Okhrana*, 1989, pp. 118–27).

although the concept of this type of unit had developed a quarter-century earlier. As the name implies, they are wildlife protection areas on which hunting is allowed. Some, such as Belovezhskaya Pushcha, Crimea, and Azov-Sivach, had been state *zapovedniki* prior to 1957, and their abrupt change in status has engendered criticism (Sokolov and Zykov, 1983), as well as suggestions to restore them to their previous status of fully protected nature reserves. There are only seven of these hunting preserves at present, and they are identified in a separate list at the end of Appendix 8.1. A great many other types of hunting lands also exist in the USSR; the most important one of these is discussed in chapter 11.

Following Stalin's death, the *zapovedniki* network again began to be expanded. Another reorganization occurred in 1961, involving reserve consolidations, the reclassification of five more as hunting preserves, and the apparent elimination of seventeen additional reserves at this time (Pryde, 1972, p. 52). The total area in reserves was decreased by approximately 35 percent as a result of this reorganization, and

Figure 8.1   View to Volga River from Zhiguli *zapovednik*. Oil development caused this nature reserve to be closed from 1951 to 1966

included the loss of two of the largest, the Altai and Kronotskiy (both since reestablished). In 1964, the area preserved in state *zapovedniki* was barely larger than it had been in 1933.

Since 1961, the system has steadily expanded without further disruptions. In 1989, over 150 reserves were spread over all portions of the Soviet Union (Figure 8.2). Particularly extensive additions to the system have occurred over the past two decades in Central Asia and Siberia. Indeed, Table 8.1 shows that almost as much new territory was added to the system during the years 1981–87 as during the entire first sixty years of the Soviet period. A list of the present *zapovedniki* of the USSR (as of the start of 1989) is given in Appendix 8.1, together with their area, location, and date(s) of creation.

Current trends

Several trends have been apparent in the management and creation of Soviet *zapovedniki* during the 1970s and 1980s. One commendable trend has been a propensity toward creating very large reserves, primarily in Siberia and the Far East. This phenomenon can clearly be seen in a listing of the largest reserves of the country (Table 8.2), in

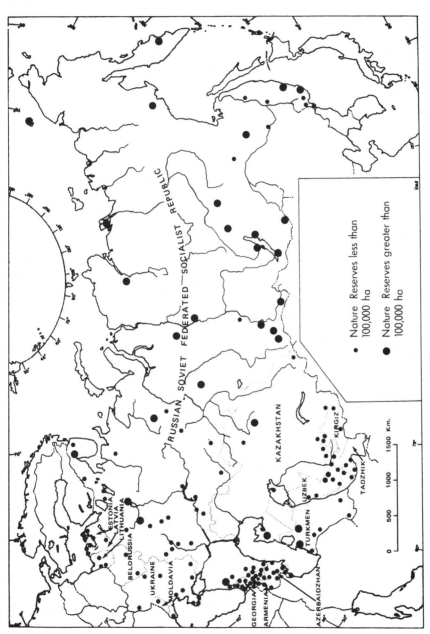

Figure 8.2 Distribution of *zapovedniki* in the USSR

Table 8.2   *Largest zapovedniki of the Soviet Union*

Reserve name	Year est.	Union republic	Area (ha)
Ust'-Lenskiy	1986	Russian	1,433,000
Taimyr	1979	Russian	1,348,316
Tsentral'no-Sibir	1985	Russian	972,017
Kronotskiy	1967	Russian	964,000
Magadan	1982	Russian	869,200
Altai	1968	Russian	863,728
Olekhminsk	1984	Russian	847,102
Wrangel Island	1976	Russian	795,650
Pechora-Ilych	1930	Russian	721,322
Baikalo-Lena	1986	Russian	659,919
Yugan	1982	Russian	648,636
Verkhne-Taz	1986	Russian	631,308
Vitim	1982	Russian	585,021
Kaplankyr	1979	Turkmen	570,000
Sayan-Shushenskoye	1976	Russian	389,570
Bureya	1985	Russian	350,000
Sikhote-Alin	1935	Russian	347,532
Azas	1985	Russian	337,300

Note: All *zapovedniki* listed above in the Russian Republic (RSFSR) are located east of the Ural Mountains except Pechora-Illych.

which fourteen of the eighteen largest that existed in 1987 are indicated as having been established since 1976, with sixteen of the eighteen located in the Russian Republic (RSFSR) east of the Ural Mountains. As noted earlier, the size and remoteness of many of these reserves results in their serving in large part a *de facto* wilderness function.

Unfortunately, many important natural areas in the Soviet Union are unrepresented by large nature reserves. Only one of these eighteen large *zapovedniki* exists in the European portion of the country (although four of between 100,000 and 300,000 ha are found there), and only one is located in Central Asia. The one large reserve in the European USSR is the Pechora-Ilych, located just west of the northern Urals, and ironically was the first of the eighteen to be created. More expansive reserves in the European part of the USSR would be particularly desirable, as this is an area where the generally more intensive scale of economic activities has made biotic conservation an issue of considerable concern. However, the very extensiveness of this economic transformation itself generally precludes the availability of large candidate sites other than in the Arctic.

Table 8.3 *Quantitative trends in the creation of new zapovedniki*

Period	Number of years	Number of new reserves	Average number created per year	Total ha at end of period	Average ha added per year	Average ha added per reserve	Number of reserves at start of period	Number of reserves at end of period[a]
1937–50	14	91	6.5	c. 12,500,000	387,000	58,900	37	128[a]
1951–60	10	53	5.3	6,360,000	489,000	92,300	40[a]	93[a]
1961–66	6	10	1.7	c. 4,700,000	57,600	34,600	62[a]	72[a]
1967–76	10	35	3.5	8,960,000	426,000	121,700	72	107
1977–86	10	42	4.2	18,742,335	978,200	232,900	107	149

Note: [a] Major reductions in the system of reserves were carried out in 1951 and 1961 (see text)
Source: Compiled by author

On the other hand, a marked increase of around fifteen in the number of *zapovedniki* in Central Asia has occurred since the mid-1970s (Gunin and Neronov, 1985). This is a propitious development since the Central Asian region is also undergoing rapid transformation. A comparison of the expansion of the reserve system by major periods, in terms of average size of reserve, is presented in Table 8.3. The last line of data in this table (1977–86) reflects the tremendous increase in average size of the *zapovedniki* created in recent years in Asiatic Russia.

A second trend is an increased concern for the conservation of rare and endangered species of flora and fauna. An overall summation of Soviet efforts of behalf of endangered species will be given in chapter 10. Here, the role of the *zapovedniki* in this regard will be summarized only briefly.

Several *zapovedniki* are identified with specific preservation efforts, some in an active, and some in a passive manner. With regard to the latter, certain preserves have been established, at least in part, to protect the nesting, denning, or wintering areas of particular species. For example, the Wrangel Island (Ostrov Vrangelya) reserve protects polar bear denning areas, and the large reserve on the Taimyr Peninsula includes the major nesting area of the red-breasted goose. Active research on behalf of establishing endangered species is carried out at such *zapovedniki* as Oka (cranes), Kavkaz (European bison), Berezina (beaver), and Kedrovaya Pad (Amur leopard). A representative (but necessarily incomplete) list of key *zapovedniki* and the specific species of rare or endangered fauna with which they are associated is presented in Table 8.4.

Since the late 1970s, several *zapovedniki* have been designated as world Biosphere Reserves, under the United Nations' "Man and the Biosphere" program (Sokolov, 1981). The purpose of this program is to establish a worldwide network of protected areas devoted to a combination of biosphere preservation and scientific research; indeed, Soviet specialists have noted that most units of their *zapovedniki* system intrinsically meet this definition. To assist the goal of biotic preservation on the biosphere reserves, some have had protective buffer zones created around them. For example, the Central Chernozem reserve, which encompasses, 4,847 ha, is surrounded by buffer zones totalling another 9,110 ha (Figure 8.3). The Oka Biosphere Reserve has been the site of considerable joint US–USSR activity on behalf of establishing common standards and procedures for more effective biosphere monitoring.

Initially seven reserves were so designated (Pryde, 1984b), but by

Table 8.4   *Zapovedniki involved with preserving rare or endangered fauna (partial list)*

Preserve	Location[a]	Ha	Species
Alma-Ata	Kazakh SSR	91,552	Snow leopard
Askaniya-Nova	Ukrainian SSR	11,054	Great bustard, kulan
Astrakhan'	Astrakhan' oblast	63,400	Sturgeon, European pelican
Badkhyz	Turkmen SSR	88,028	Kulan, gazelle
Barsa-Kelmes	Kazakh SSR	18,300	Gazelle, bustard
Barguzin	Buryat ASSR	263,176	Sable
Berezina	Byelorussian SSR	76,201	River beaver
Kavkaz	Krasnodar krai	263,477	European bison
Kedrovaya Pad'	Primorskiy krai	17,897	Amur leopard
Kopet-dag	Turkmen SSR	49,793	Mountain sheep
Lazov	Primorskiy krai	116,524	Goral, Siberian tiger
Lugansk	Ukrainian SSR	1,580	Steppe marmot
Mordov	Mordov ASSR	32,148	Desman
Naurzum	Kazakh SSR	87,694	Little bustard
Nuratinsk	Uzbek SSR	22,537	Lammergaier
Oka	Ryazan oblast	22,896	Siberian crane, desman
Repetek	Turkmen SSR	34,600	Desert sparrow
Sikhote-Alin	Primorskiy krai	347,532	Siberian tiger, goral
Taimyr	Krasnoyarsk krai	1,348,316	Red-breasted goose
Teberda	Stavropol krai	84,996	Caucasian black grouse
Tigravaya balka	Tadzhik SSR	47,409	Bukhar deer, gazelle
Wrangel Island	in Arctic Ocean	795,650	Polar bear, snow goose

Note: [a]   Republic (SSR) in which it is located, or, if in the RSFSR, the province (oblast) or autonomous republic (ASSR) in which it is found
Sources: Pryde, 1972; Kolosov, 1982; Borodin and Syroyechkovskiy, 1983; Borodin (ed.), 1985; Knystautas, 1987

the late 1980s the total had risen to at least nineteen (Table 8.5). A disproportionate number now exist in the Russian Republic, however. It has been suggested that perhaps another ten might be created in the near future, with such reserves as the Altai, Badkhyz, Carpathian, Darwin, Oka, Zakataly, and others being good candidates (Krinitskiy, 1981). Ultimately, a network of as many as 305 biosphere reserves, based on the many distinctive physiographic regions of the USSR, have been proposed (Gavva *et al.*, 1983). The first International Biosphere Reserve Congress was convened in Minsk in September of 1983.

Another contemporary trend is an emphasis on more sophisticated forms of environmental monitoring. Although all of the reserves keep records of the natural conditions and seasonal progressions that occur

Table 8.5   *Soviet biosphere reserves (in 1988)*

Name	Republic	Est'd. [a]	Area (ha)
Askaniya-Nova	Ukraine	1921, 1985	11,054
Astrakhan	Russian	1919, 1984	63,400
Berezina	Belorussia	1925, 1978	76,201
Central Chernozem (Tsentralnochernozemnyy)	Russian	1935, 1978	4,795
Central Forest (Tsentralnolesnoy)	Russian	1931, 1985	21,348
Central Siberian (Tsentralnosibir)	Russian	1985	972,017
Chatkal Mountains [b]	Uzbek, Kirgiz	1947, 1978	71,400
Chernomor'ye	Ukraine	1927, 1984	71,899
Kavkaz (Caucasus)	Russian	1924, 1978	263,477
Kronotskiy	Russian	1967, 1984	964,000
Lake Baikal [c]	Russian	1916, 1984	559,100
Lapland	Russian	1930, 1984	161,254
Oka River [d] (Prioksko-Terrasnyy)	Russian	1945, 1978	45,845
Pechora-Ilych	Russian	1930, 1984	721,322
Repetek	Turkmen	1928, 1978	34,600
Sayan-Shushenskoye	Russian	1976, 1984	389,570
Sikhote-Alin	Russian	1935, 1978	347,532
Sokhonda	Russian	1973, 1984	210,986
Voronezh	Russian	1927, 1984	31,053

Notes: [a] The first date shown is when the *zapovednik* was established; the second date is the year it was given biosphere reserve status
[b] Originally was the Sary-Chelek *zapovednik*; now also includes the Chatkal *zapovednik* in the Uzbek republic
[c] Originally was the Barguzin *zapovednik*; now also includes the Baikal *zapovednik* and other territories
[d] Originally was only the Oka Terrace (Prioksko-terrasnyy) *zapovednik*; the present biosphere reserve takes in a much larger area

within them, today many reserves are monitoring such variables as groundwater levels and composition, the quality (including pH and harmful constituents) of water and air, and changes to soil and biota that can be attributed to external activities. This is especially true of the biosphere reserves. Some reserves, such as the Central Chernozem, have long-established research stations operated by units of the Academy of Sciences and are well organized for this sort of complex monitoring (Grin, 1982; Figure 8.4). The topic of environmental monitoring will be reviewed in more detail in chapter 13.

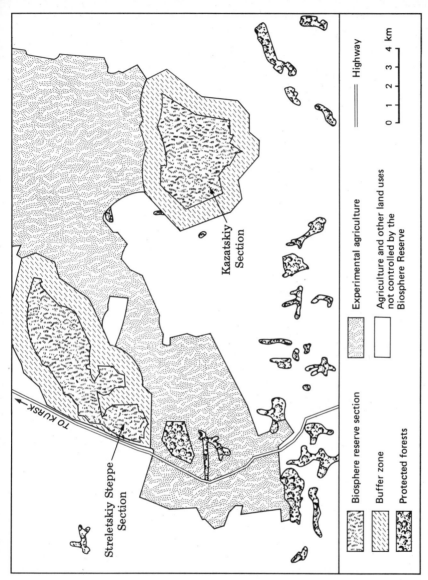

Figure 8.3 Central Chernozem Biosphere Reserve, with buffer zone

Biosphere reserve section

Buffer zone

Protected forests

Experimental agriculture

Agriculture and other land uses
not controlled by the
Biosphere Reserve

—— Highway

0 1 2 3 4 km

TO KURSK →

Streletskiy Steppe
Section

Kazatskiy
Section

Figure 8.4 Soil research site at Central Chernozem Biosphere Reserve

Finally, in reflection of the increasing demand for organized recreational opportunities in the Soviet Union, and the fact that *zapovedniki* are not intended to fulfill this role, the network of nature reserves is now being supplemented by a new network of national parks. Public recreation is the main, but not the only, purpose of the new national parks, and they will be examined in more detail in chapter 9.

Management problems

Preserved areas anywhere in the world are constantly under assault from a variety of problems. The nature of the problems will vary from one site to another, but all require continuing attention if the integrity of the preserve is not to be compromised. Most commonly, *zapovedniki* find themselves periodically vulnerable to competing economic interests. As noted above, this originated in the anti-preservation, utilitarian philosophies of the Stalin era, when such activities as logging in the Central Forest reserve and exotic game breeding at Askania-Nova were not only permitted but encouraged. This culminated in the massive assault on the system in 1951. Today, the assaults on *zapovedniki* integrity generally originate from within the developmental aspirations of some republic or all-Union ministry.

A common threat to preserved areas are outside sources of pollution. At the Central Chernozem biosphere reserve, for example, which is located in a region of extensive agriculture, monitoring pesticide residues and other atmospheric pollutants is a major area of research. The monitoring is carried out both in a buffer zone around the reserve, and on agricultural test plots within the *zapovedniki* itself (Pryde, 1984b). At reserves along the Amu-Darya river, both pesticides and nutrients from irrigated agriculture, exacerbated by flow reduction due to irrigation water diversions, threaten the biological integrity of the units.

Several additional reserves in the European part of the USSR face similar problems from surrounding economic activities. At Oka-Terrace, which is also a biosphere reserve, many important nearby wetland areas ideally should be added to the *zapovedniki* to preserve them. This would quadruple its size (Mekayev, 1981). The same author reports serious incursions into the buffer zones around some of the isolated reserves in the Ukrainian steppe. In the Volga delta, a gas refinery was shut down after its excessive emissions posed harm to both farms in the vicinity and the Astrakhan biosphere reserve (*Pravda*, June 17, 1989).

To illustrate the range of potential external pollution problems, one hunting preserve, the Dneprovsko-Teterevskoye in the Ukraine, lay partially within the evacuation zone surrounding the Chernobyl nuclear explosion site (Figure 3.2). Although no specific results of radioactive monitoring are available from this hunting preserve, it was placed on the International Union for the Conservation of Nature (IUCN) world list of threatened preserves (*IUCN Bulletin*, 1986, p. 132). In addition, the Polesskiy and Pripyat *zapovedniki* lay down wind from the smoke plume from the crippled reactor, and probably received a significant amount of fallout.

Examples of ministerial threats to the *zapovedniki* abound. In the 1970s, it was proposed to build a road through the Caucasus mountains that would have threatened the Kavkaz reserve. The road has not yet been built, but serious problems of poaching in the same preserve have subsequently been reported (Volkov, 1980). A proposed hydro-electric project in Armenia in the mid-1980s that would have harmed the Khosrov reserve was required to be redesigned (Bablumyan, 1986). An ongoing fight is taking place in the Uzbek republic between the Tigrovaya Balka reserve and various economic interests who would use some of its territory for agricultural pursuits (Illesh and Surkov, 1985). The newspaper *Pravda Ukrainy* in 1987 carried a feature article on what it felt was excessive wood cutting in the Karpatskiy (Carpathian) *zapovedniki* (Komendar, 1987). Other examples have included poaching at Kyzyl-Agach, logging around Sikhote-Alin, hay harvesting within Sary-Chelek, and air pollution damage at Il'men (Komarov, 1980). Due to the drying up of the Aral Sea (chapter 12), the insular preserve of Barsa-Khelmes could eventually connect to the mainland, with significant ecological implications. The cases cited above indicate that only on occasion are such threats to the *zapovedniki* averted (often assisted by local public outcry), and that continual vigilance would seem to be a prerequisite to long-term nature preserve integrity in the Soviet Union, just as it is elsewhere.

Although it is always stressed that *zapovedniki* are not to play a major tourism role, it has been noted that in fact a few still do. It would seem in keeping with the stated management intentions for *zapovedniki* that either the tourist operations should be terminated at these reserves, or else they should be given some other designation, most likely national parks. Although the latter option would seem a relatively easy step to take, to date neither of the above courses of action has been adopted at the tourist-accommodating *zapovedniki* such as Teberda, Ritsa, and Stolby. In addition, new tourist facilities are being built in the vicinity of the reserve at Lake Issyk-Kul', and it is starting to take on the appearance of a national park, as well (Gavva *et al.*, 1983). This lake has also suffered from pollution and from loss of water due to irrigation diversions.

For over twenty years it has been repeatedly suggested by Soviet specialists that a unified administrative system is needed to manage the network of reserves (Alekseyeva *et al.*, 1983; Knystautas, 1987; Mekayev, 1981; Volkov, 1977, and many others). At present, a variety of organizations, some of them identified at the start of this chapter, are involved with administering them; there is no equivalent of the National Park Service in the United States to provide uniform management policies and procedures for the system. As a result, it is difficult even to get an "official" and accurate list of existing reserves. Here is one instance where the creation of a new governmental agency might clearly be defensible. In the absence of a "State Zapovedniki Service," the new State Committee for Environmental Protection (Goskompriroda) appears to have been designated as such an agency. Among its many duties, it has been charged with "managing nature reserves" (*Pravda*, Jan. 17, 1988). But it is not clear whether this refers only to those that have been brought under its jurisdiction (120 as of 1989; Doklad ..., 1989, p. 132), or whether all of them are intended to be subsumed under its management. The question of what form of "management," if any, it will exercise over the remainder is also unspecified in the 1990 draft law on preserved lands (Zakon, 1990).

Of considerable importance is the overall size and inclusiveness of the *zapovedniki* network. In terms of size, even with the recent large additions, only about one-half of 1 percent of the Soviet Union is represented by the reserves (Ryzhikov, 1988). Although other types of preserved areas exist, they are either relatively small in area (the national parks take in less than a tenth the area of the *zapovedniki*), or, like the hunting preserves, offer a lesser degree of protection. Even if all of them are considered, the total percentage of the USSR in such units is only about 2 percent, well under recommended norms.

The term "inclusiveness" refers to how well all the various natural zones of the USSR are represented in the system (Isachenko, 1989). For example, there are presently no biosphere reserves at all in the biologically diverse Transcaucasus region. The huge new reserves in Siberia and the Far East are commendable, but biotic diversity is both greater and more threatened in several other natural zones, such as all steppe regions, riparian corridors and other wetlands, and some remaining portions of Central Asia. Because of the scale of existing development, new reserves in these areas may not be able to be large in size, but even smaller areas would be helpful and potential sites appear to exist (Mekayev, 1981). But trying to establish new reserves in some of the smallest of the remnant natural habitats may not be sufficient; to be viable, reserves must be of at least a certain minimum size (which will be variable depending on the habitat involved). Thus, connecting isolated sites by incorporating both them and the land between them into new reserves, even if this means withdrawing some land from agricultural development and spending sizable sums on restoration, may be necessary in critical habitats such as steppe, coastal, and riparian areas.

Finally, the *zapovedniki* system suffers from a disease common to every preserved natural area in the world: inadequate funding. A great disparity exsits (up to a factor of almost twenty times) in the average amount of annual funding received by reserves in various parts of the country (Ryzhikov, 1988). Some of the more recently created reserves have no permanent on-site research bases, facilities, or staff available to them, and those that do are often constrained in the scope of their activities. In some cases, additional funds are needed for increased enforcement capabilities, especially against poaching. Equally as important as money is the perceptual priority given to the *zapovedniki* system by the government. In the intense ministerial battle for budgetary funds, it is easy to view the *zapovedniki* as not the highest priority, especially since they lack a unified managerial voice to speak for them.

Despite the above problems, the Soviet *zapovedniki* system is recognized as one of the world's most important nature protection networks. Its current rapid expansion bodes well for the protection of at least some major components of Soviet flora and fauna. As the size of the reserve network continues to increase in the future, its significance for biosphere protection will likewise increase.

Appendix 8.1

Appendix 8.1   *1988 Zapovedniki of the USSR*

Name	SSR	1st established	Abolished	Re-established	Present ha
Adzhameti	Gru	1946	1951	1957	4,848
Ak-Gyol'	Azb	1978			9,100
Akhmetskiy	Gru	1980			16,297
Aksu-Dzhabagly	Kaz	1926			74,416
Algetskiy	Gru	1965			6,000
Alma-Ata	Kaz	1931	1951	1961	91,552
Altai	Rus	1932	1951	1968	863,728
Amudarinskiy	Tur	1982			50,506
Aral-Paigambar	Uzb	1960?	?	1971	3,094
Arnasai	Uzb	1977			63,368
Askaniya-Nova	Ukr	1874		1921	11,054
Astrakhan	Rus	1919			63,400
Azas	Rus	1985			337,300
Badai-Tugai	Uzb	1971			6,497
Badkhyz	Tur	1941			88,028
Baikal	Rus	1969			165,724
Baikalo-Lena	Rus	1986			659,919
Barguzin	Rus	1916			263,176
Barsa-Khelmes	Kaz	1939			18,300
Basegi	Rus	1982			19,422
Bashkir	Rus	1930	1951	1958	72,140
Basutchai	Azb	1974			117
Batsara-Babanauri	Gru	1935	1951	1957	3,812
Berezina	Bel	1925			76,201
Besh-Aral'skiy	Kir	1979			18,200
Bol'shekhekhtsir	Rus	1963			44,928
Borzhomi	Gru	1935	1951	1959	18,048
Bureya	Rus	1985			350,000
Chapkyalyai	Lit	1975			8,469
Chatkal	Uzb	1947			35,809
Chernomorskiy	Ukr	1927			71,899
Dagestan	Rus	1987			19,061
Dal'nevostochnyy	Rus	1978			64,360

Appendix 8.1   (*cont.*)

Darvin	Rus	1945			112,630
Dashti-Dzhum	Tad	1985			19,700
Daurskiy	Rus	1988			44,752
Dilizhan	Arm	1958			31,193
Dunaiskie Plavni	Ukr	1973		1981	14,851
Endlaskiy	Est	1985			8,162
Galich'ya Gora	Rus	1925	1951	1969	231
Geigel'	Azb	1925	1951	1965	7,131
Girkan	Azb	1936		1969	2,900
Gissar	Uzb	1984			87,500
Grini	Lat	1936	1951	1957	1,076
Gumista	Gru	1946	1951	1976	13,400
Ilmen	Rus	1920			30,380
Ismailimskiy	Azb	1981			5,778
Issyk-kul'	Kir	1948	1951	1976	17,310
Kabardino-Balkar	Rus	1976			74,081
Kamanos	Lit	1979			3,660
Kandalaksha	Rus	1932		1939	58,100
Kanev	Ukr	1931	1951	1968	1,030
Kaplankyr	Tur	1979			570,000
Karadag	Ukr	1979			1,370
Karakul'	Uzb	1971			14,331
Karayaz	Azb	1978			4,769
Karpatskiy	Ukr	1968			18,544
Kavkaz	Rus	1924			263,477
Kazbek (Kazbegi)	Gru	1976			4,300
Kedrovaya Pad	Rus	1916			17,897
Khingan	Rus	1963			82,186
Khopyor	Rus	1935			16,178
Khosrov (Garni)	Arm	1958			23,425
Kintrishi	Gru	1959			7,166
Kitabskiy	Uzb	1980			5,378
Kivach	Rus	1931			10,460
Kodry	Mol	1971			5,159
Kolkhid (Colchis)	Gru	1935	1951	1959	500
Komsomol'sk	Rus	1963			61,208
Kopetdag	Tur	1976			49,793
Kostomuksh	Rus	1983			47,569
Krasnovodsk (Gasan-Kuli)	Tur	1932		1968	262,037
Kronotskiy	Rus	1934	1961	1967	964,000
Krustkalny	Lat	1977			2,826
Kurgal'dzhino	Kaz	1959	1961	1968	237,138
Kurile	Rus	1984			63,365
Kyzyl-Agach	Azb	1929			88,360
Kyzyl-Kum	Uzb	1971			3,985
Kyzylsu (= Gissar since 1984)	Uzb	1975			0
Lagodekhi	Gru	1912			17,818
Lapland	Rus	1930			161,254
Lazov (Sudzukhe)	Rus	1935	1951	1957	116,524

Appendix 8.1   (*cont.*)

Les na Vorskle	Rus	1925	1979	1,038
Liakhvi	Gru	1977		6,804
Lugansk	Ukr	1968		1,580
Magadan	Rus	1982		869,200
Malaya Sos'va	Rus	1976		92,921
Mariamdzhvari	Gru	1935	1951 1959	1,040
Markakol'	Kaz	1976		71,367
Matsalu	Est	1957		48,634
Mirakinsk (= Gissar since 1984)	Uzb	1976		0
Mordov	Rus	1935		32,148
Moritssala	Lat	1912	1957	818
Mys Mart'yan	Ukr	1973		240
Naryn	Kir	1984		48,000
Naurzum	Kaz	1931	1951 1965	87,694
Nigula	Est	1957		2,771
Nizhne-Svir	Rus	1980		41,000
Nuratinsk	UZB	1975		22,537
Oka	Rus	1935		22,896
Olekhminsk	Rus	1984		847,102
Ostrov Vrangelya (Wrangel Island)	Rus	1976		795,650
Pechora-Ilych	Rus	1930		721,322
Pinega	Rus	1975		41,244
Pirkuli	Azb	1968		1,521
Pitsunda-Myussera	Gru	1926	1951 1957	3,770
Polesskiy	Ukr	1968		20,104
Prioksko-Terrasnyy	Rus	1948		4,945
Pripyat'	Bel	1969		62,213
Pskhu (Pskhuskiy)	Gru	1978		27,643
Ramit	Tad	1959		16,139
Rastoch'ye	Ukr	1984		2,080
Repetek	Tur	1928		34,600
Ritsa	Gru	1946	1951 1957	16,289
Saguramo	Gru	1946	1951 1957	5,247
Sary-Chelek	Kir	1959		23,868
Sataplia	Gru	1935	1951 1957	354
Sayan	Rus	1983		120.000
Sayano-Shushenskiy	Rus	1976		389,570
Severo-Osetinsk	Rus	1967		25,903
Shirvan	Azb	1969		17,745
Shul'gan-Tash	Rus	1986		22,531
Sikhote-Alin	Rus	1935		347,532
Slitere	Lat	1921	1957	14,882
Sokhonda	Rus	1973		210,986
Stolby	Rus	1925		47,154
Syunt-Khasardag	Tur	1979		29,700
Taimyr	Rus	1979		1,348,316
Teberda	Rus	1936		84,996
Teychi	Lat	1982		18,966
Tigrovaya Balka	Tad	1938		47,409

Appendix 8.1   *(cont.)*

Tsentral'nochernozem	Rus	1935		4,795
Tsentral'nolesnoy	Rus	1931	1951	1960	21,348
Tsentral'nosibir	Rus	1985		972,017
Turianchai	Azb	1958		12,356
Ukrainskiy Stepnoi	Ukr	1926	1961	1,634
Ussuri (Suputinka)	Rus	1911	1932	40,432
Ust'-Lenskiy	Rus	1986		1,433,000
Ust'-Yurt	Kaz	1985		223,000
Vardanzinsk	Uzb	1973		324
Vashlovani	Gru	1935	1951	1957	4,868
Verkhne-Taz	Rus	1986		631,308
Vil'sandi (Vayka)	Est	1910	1958	10,689
Visim	Rus	1946	1951	1971	13,750
Vitim	Rus	1982		585,021
Viydumyae	Est	1957		593
Volga-Kama	Rus	1960		8,034
Voronezh	Rus	1927		31,053
Yalta	Ukr	1973		14,591
Yugan	Rus	1982		648,636
Zaamin	Uzb	1959		10,560
Zakataly	Azb	1929		25,190
Zavidovo	Rus	1972		125,442
Zeravshan	Uzb	1975		2,360
Zeya	Rus	1963		82,567
Zhiguli	Rus	1927	1951	1966	23,103
Zhuvintas	Lit	1946		5,428

Total 18,886,223
Count   155

Note: Arm: Armenia; Azb: Azerbaidzhan; Bel: Belorussia; Est: Estonia; Gru: Georgia; Kaz: Kazakhstan; Kir: Kirgizia; Lat: Latvia; Lit: Lithuania; Mol: Moldavia; Rus: Russia; Tad: Tadzhikstan; Tur: Turkmenistan; Ukr: Ukraine; Uzb: Uzbekistan

USSR Hunting reserves

Name	SSR	Pr. est.	Pr. ha
Azov-Sivach HR	Ukr	1957	57,430
Belovezhskaya-Pushcha HR	Bel	1940	87,577
Dneprovsko-Teterevskoye HR	Ukr	1968	37,891
Krym (Crimean) HR	Ukr	1957	42,957
Redenskiy Les HR	Mol	1976	5,525
Telekhany HR	Bel	1977	10,947
Zales'ye (Zalesskoye) HR	Ukr	1957	35,089
Total			277,416

9 National parks: a new feature on the Soviet landscape

National parks represent a relatively new component on the cultural landscape of the Soviet Union, one that did not exist until the early 1970s. To date, they have received relatively little discussion in the west. This chapter will summarize the nature of national parks in the Soviet Union, and identify the ways in which they are similar to, and differ from, national parks in North America.

As noted in the previous chapter, the *zapovednik* concept does not encourage tourism, and hence other types of specially designated areas were sought for this purpose. The worldwide concept of national parks was an obvious option for consideration. They constitute the Soviet Union's "open" nature reserves, areas that are partially preserved while at the same time being largely accessible to the recreational user.

History of the national park system

National parks were authorized in the USSR by the same act, signed by Lenin in 1921, that authorized the *zapovedniki* system (Pryde, 1972, p. 213). Despite much public discussion and numerous recommendations, however, the first national park was not created in the USSR until fifty years later. By 1980 there were eight such areas, and in 1988, at least eighteen. Each of the fifteen Union republics, excepting Azerbaidzhan, Belorussia, Moldavia, and the Tadzhik and Turkmen republics, had at least one by the end of 1987 (Table 9.1). Of the ten new parks created between 1983 and 1986, eight were in the Russian Republic (RSFSR), the largest of the USSR's constituent republics. Despite considerable discussion, none had been created in the RSFSR prior to 1983.

Although no national park was actually established in the USSR until 1971, they had been proposed and debated for a number of years prior to that time. The process of academicians and others suggesting

157

Table 9.1   *National parks of the Soviet Union*

	Year	Name, republic [a]	Area (ha)
1.	1971	Lahemaa NP (Estonia)	64,911
2.	1971	Gauya NP (Latvia)	83,750
3.	1971	Lithuanian NP	30,000
4.	1973	Tbilisi NP (Georgia)	19,400
5.	1976	Ala-Archa SNP (Kirgiz)	19,400
6.	1978	Sevan NP (Armenia)	150,000
7.	1978	Uzbek People's Park	32,400
8.	1980	Karpatskiy (Carpathian) SNNP (Ukraine)	50,303
9.	1983	Losiniy ostrov (Elk Island) SNNP (RSFSR)	10,067
10.	1983	Shatskiy SNNP (Ukraine)	82,500
11.	1983	Sochi SNNP (RSFSR)	190,000
12.	1984	Samarskaya Luka SNNP (RSFSR)	134,000
13.	1985	Bayanaul'skiy SNNP (Kazakhstan)	45,500
14.	1985	Mariy Chodra SNNP (RSFSR)	36,600
15.	1986	Pribaikal'skiy SNNP (RSFSR)	136,000
16.	1986	Bashkiriya SNNP (RSFSR)	79,800
17.	1986	Zabaikal'skiy SNNP (RSFSR)	209,000
18.	1986	Priel'brus'ye SNNP (RSFSR)	100,400
19.	1987	Kurshskaya kosa NP (SNNP) (RSFSR)	6,100

Notes: [a] NP = National Park
 SNP = State Natural Park
 SNNP = State Natural National Park
 RSFSR = Russian Soviet Federated Socialist Republic (the "Russian Republic")
Sources: Nikolayevskiy, 1985; IUCN, 1989; *Okhrana* . . ., 1989, p. 127

locations for national parks began at least in the mid-1960s (Belousova, 1967), however, few of the early siting suggestions have been realized in the parks actually created to date. Nevertheless, the periodic appearance of new lists of hoped-for national parks continues on to the present (Nikolayevskiy, 1985, chapter 3).

Two prospective national parks in particular seemed to be of high priority in the 1960s, one south of Moscow (referred to as "Russkiye Les"), and one along the shores of Lake Baikal. Indeed, the "creation" of both of these parks was announced in the late 1960s in both *Soviet* and western periodicals. Such announcements appeared, for example, in the referenced articles by Bannikov, 1968; Pryde, 1967; in *American Forests* (Jan. 1967); in *Pravda* (June 19, 1970), and in numerous other sources. So imminent did these two parks appear to be in 1970 that the author devoted over three pages of *Conservation in the Soviet Union* to this topic, even though no Soviet national parks had yet been created (Pryde, 1972, pp. 62–5). Further, in 1974, *Izvestiya* announced the approval of a plan for the first of five national parks around Lake Baikal alone (June 17). However, as Soviet authors frequently lament, no park was established at Lake Baikal until 1986 (Isakov, 1979; Reimers, 1982).

The current status of the "Russkiye Les" is unclear. The October, 1976, issue of the Soviet journal *Turist* refers to an existing development in this area as an "experimental-demonstration forestry complex," and elsewhere it has been called the "Russian Forest Experimental and Model Farm"; apparently it is what might be termed in the United States a "demonstration forest." However, it has also been called a "logging enterprise" where timber procurement is paramount (Isakov, 1979). In the same article, Isakov offered a "tentative list" of ten initial national parks proposed for the Russian Republic, but of these only four are in or near locations where parks have actually been created to date.

In 1971, the first official Soviet national park did appear, in the Estonian republic. Called "Lahemaa," it was enacted by the Council of Ministers of Estonia on June 1 of that year. This and the other Soviet national parks will be briefly described in a subsequent section.

Although the concept of creating national parks *per se* has been generally popular in the Soviet Union, there has been less unanimity over what they should be called. Table 9.1 indicates that in fact four different designations are currently used for such parks in the USSR: National Park, People's Park, State National Park, and State Natural National Park (*gosudarstvennyy prirodnyy natsional'nyy park*).

The preference seems to be determined by the particular Union republic involved: all nine of the RSFSR parks, as well as those in the Ukraine, are called state natural national parks, whereas in the Baltic and Transcaucasus republics the simpler "national park" seems to be preferred. Only the Kirgiz park "Ala-Archa" uses the designation "state natural park," and only Uzbekistan opts for "People's Park."

The use of the term "natural park" in the names of many of these areas may have originated from a discussion dating from 1968, in which a leading Soviet conservationist argued for this appelation (Bannikov, 1968). His argument was that the term "national" park, used in the west to describe federally owned and managed parks, was inappropriate in the USSR since all land was nationalized (and perhaps because the Soviet parks were to be created by the Union republics rather than by a USSR agency). He suggested the use of "natural park" instead. The preferred adjective to convey the idea of an area being "national" in importance, an adjective which is also used in conjunction with *zapovedniki*, seems to be "state" (*gosudarstvennyy*), at least in the slavic republics. The rather cumbersome "state natural national park" designation may be nothing more than a typical bureaucratic compromise.

Management policies and problems

Discussions of the appropriate management principles for Soviet national parks began in the 1960s (Bannikov, 1968; Belousova, 1967; Borisov, 1968). The first Soviet national park, "Lahemaa" in Estonia, in many ways became a model for those that followed. The national parks are seen as complimenting the long-established network of nature reserves, or *zapovedniki*, which were noted in the previous chapter as not being intended to serve a tourist function.

The national parks are created by legislative acts of the individual Union republics in which they are located. As a result, there exists neither a national agency to manage them, nor a uniform set of national management policies, a situation similar to that existing in the case of the *zapovedniki*. The Academy of Sciences' Institute of Geography has made recommendations as to how Soviet national parks should be sited and managed, including studies on carrying capacities (Gerasimov and Preobrezhenskiy, 1983), as have others.

Despite this bureaucratic fragmentation, however, a certain similarity among them in concept can be observed. All of the early parks embrace the spatial design concept of including quasi-wilderness (restricted) zones, intensive recreation zones, and zones of economic activity (Table 9.2). In this way they are similar to national parks in North America, which are divided into wilderness and non-wilderness sections, with the wilderness portion usually being the larger. However, Table 9.2 indicates that in the early Soviet parks, the restricted zone only averages 9 percent of the total park area.

In certain other respects, though, the Soviet parks differ markedly from their American counterparts; many more resemble the British model of "landscape" parks. For example, there is no systematic effort to include the country's outstanding geological features in the assemblage of parks. Rather, the primary goal of most appears to be the preservation of one or more of the typical landscapes (both natural and cultural) of the republic or province in which they are located. Also differing from both American and British practice, one (Losiniy ostrov) lies largely within the city limits of Moscow. In some respects, they

Table 9.2 *Comparison of five of the earlier national parks*

Republic	Lahemaa Estonia	Gauya Latvia	Lithuanian Lithuania	Sevan Armenia	Carpathian Ukraine	Average
Area (ha)	65,000	84,000	30,000	150,000	50,000	76,000
Restricted zone, 1,000 ha/%	4/6	4/5	1/3	15/10	10/20	9%
Protected recreation zone, 1,000 ha/%	35/54	33/39	14/47	120/80	21/42	52%
Intensive recreation zone, 1,000 ha/%	4/6	2/2	2/6	5/3	11/22	8%
Zone of economic activities, 1,000 ha/%	22/34	45/54	13/44	10/7	8/16	31%
Average visitation annually	100,000	1,000,000	70,000	200,000	500,000	374,000
Annual budget, rubles	530,000	630,000	85,000	450,000	700,000	479,000

Source: Adapted from Nikolayevskiy, 1985, p. 97

perhaps more closely resemble national recreation areas than national parks in the United States.

The lack of a unified management agency has come under occasional criticism. One specialist laments that "each park is accountable only unto itself. There are no general legal principles, no economic status, and no rules on staffing . . . In short, both the parks and the tourist routes need a single master" (Reimers, 1980). The same writer in a later article suggests that inadequate funding is being made available, and also urges that the national park concept be extended to include the creation of national historical parks (Reimers, 1982).

It is clear that the management principles for Soviet national parks are still in somewhat of a formative stage. It is entirely possible that this question will always be left up to the individual Union republics, and that no such concept as an all-USSR national park administration will ever come into being. If the latter course of action is taken, however, it can be hoped that a unifying administrative agency will appear at least in the large Russian republic, and with time (and the creation of additional parks) in the other fourteen republics as well.

The first decade of Soviet national parks

During the first ten years of the national park era in the USSR, eight such parks were created. The first three appeared in the Baltic republics, followed by two in the Transcaucasus, two in Central Asia, and one in the Ukraine (Figure 9.1).

The first park in the Soviet Union was Lahemaa in Estonia, taking in 64,911 ha on the Gulf of Finland (Eilart, 1976). The goals of the park are (1) to preserve and popularize the natural and cultural heritage; (2) to conduct scientific research; and (3) to increase public awareness of environmental problems (Kaasik and Kask, 1983). Descriptive literature on the park indicates that it has been divided into the following types of management zones: reserve areas, which are restricted research areas off limits to tourists (6–8 percent of the total area); landscape preservation districts where only hiking and other passive forms of recreation are permitted (about two-thirds of the total area); and various "cultivated landscapes" (30 percent of the total). The latter includes moderate recreation areas where camping and other active pursuits are allowed, intensive recreation areas which have been created around existing settlements, active forestry areas, and, on about 5 percent of the park, parcels on which agricultural pursuits are permitted (Lakhemaa National Park, 1975).

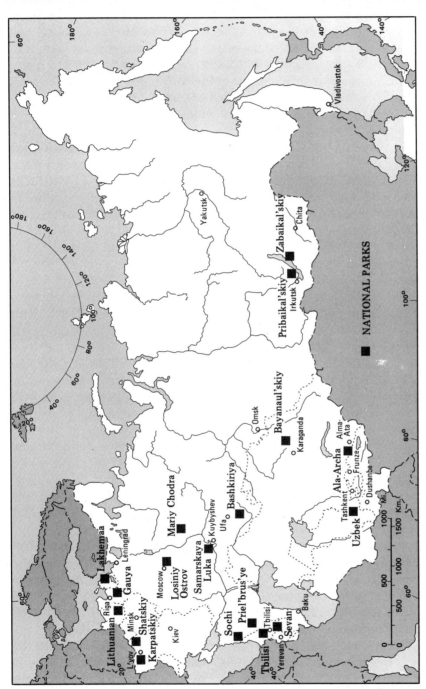

Figure 9.1 Location of Soviet national parks

The park contains numerous archaeological sites, and several manorial estates. One of the latter, the Von Pahlen estate at Palmse, will eventually serve as one of the main visitor and interpretive centers of the park. Park officials indicated to the author that funding is a major problem, causing development of the park to proceed at a very slow pace. They have attempted to solicit private donations on behalf of the park, but this effort has only been modestly successful.

In the same year that Lahemaa was created (1971), Gauya National Park was established in Latvia. The largest of the Baltic parks, its area is usually given as 83,750 ha, although a 1988 park brochure gives the size as 92,048 ha. It is located 50 km from Riga along the forested hills and valleys of the Gauya River and its tributaries. The three zones in the park are described as conservation zones, recreation zones, and "neutral" zones; the first two are further divided into sub-units of varying intensities of use, the latter includes agricultural activities and villages. Historical preservation is a dominant theme of the park; it contains castles dating back to the thirteenth century and hundreds of archeological and cultural sites (Figure 9.2). The most intensive recreation zones are near the settlements of Sigulda and Tsesis, with Sigulda receiving as many as 900,000 vacationers a year (Bergholtsas, 1976).

Also created in 1971 was the Lithuanian National Park of 30,000 ha. It is located in the republic's lake district, immediately west of the city of Ignalina. The park's sixty-two lakes take in 16 percent of its area (*Eesti loodus*, May 1975). Its three primary zones are scientific-conservation, recreation, and agriculture-forestry. Each of these is divided into various administrative sub-zones. It is also surrounded by a buffer zone, the area of which is not included in the park's stated hectarage. The park has been well developed with diverse recreational opportunities. In the same general area as the park is the Ignalina Atomic Power Station, one of the largest nuclear generating complexes in the USSR (see chapter 3).

Two different dates are given for the establishment of the fourth Soviet national park, Tbilisi in the Georgian republic. It was initially created in 1973 under the Georgian Ministry of Forestry, but it was not accepted onto the official list of USSR national parks until 1984 (Nikolayevskiy, 1985). It is a relatively small park, taking in 19,400 ha, and is located not far from the capital city of Tbilisi.

Ala-Archa in Kirgizia was the first park in Central Asia, and is the only one to bear the title of a "state natural park." Like the Tbilisi park, it covers 19,400 ha. Located in mountainous country along the upper

Figure 9.2   View from Turaida castle in Gauya National Park

Ala-Archa River 30 km south of Frunze, it is situated to serve the recreation needs of the residents of the capital city. However, the extent of its development for these purposes has been criticized (*Sovetskaya Kirgizia*, Aug. 14, 1986).

The largest of the early parks is Sevan, in Armenia. Located along the shores of the high lake of the same name, it encompasses 150,000 ha. Unfortunately, water diversion projects have greatly reduced the size of Lake Sevan, to the point where a succulent lake trout called the *ishkhan* is threatened. Although measures to raise the lake's level by 8 m are under way, so as to restore the earlier natural balance, this will not be completed prior to the year 2000 (*Gudok*, Sept. 16, 1981).

The second national park in Central Asia was the Uzbek People's

Park, also the only one using that particular designation. Created in early 1978, it covers 32,400 ha. It is located 150 km east of Samarkand in the Zaamin mountains of the Pamir-Alai range. The region of the park is predominately arid, though conifer forests are found at the higher elevations in the park, which extend up to 2,600 m (Alibekov, 1986).

In the Ukraine, the initial national park was the Carpathian (Karpatskiy) State Natural National Park, the first unit to employ this longer title. Created in 1980, its area is just over 50,000 ha. It is located in the Carpathian mountains of Ivano-Frankovskiy province; it contains peaks over 2,000 m and is covered by forests over three-quarters of its territory (Nikolayevskiy, 1985).

The new parks of the 1980s

Following the creation of the Carpathian park in 1980, no new parks were established for three years. Beginning in 1983, however, activity increased rapidly. In the next three years, ten new parks were created, eight in the Russian Republic and one each in Kazakhstan and the Ukraine. The new national parks of the 1980s averaged considerably larger in size than their predecessors. The first eight parks had an average size of 55,000 ha, whereas those created after 1980 averaged about 110,000 ha, or approximately twice as large.

The first two national parks in the Russian Republic (RSFSR), Losiniy ostrov and Sochi, were established in 1983. Losiniy ostrov ("Elk Island") is the smallest of the parks at 10,067 ha. It is located partly within the city of Moscow, and embraces what was a large section of the city's perimeter green belt. Although providing recreation for Muscovites, it was noted in the press in 1987 that economic incursions were taking place in the park. It has benefited from public volunteers who have worked to enhance its attractiveness (*National Parks*, Nov./Dec., 1988, p. 17).

The Sochi park is located along the Black Sea coast near the city of the same name, and extends inland to the foothills of the Caucasus mountains. With 190,000 ha, it is the Soviet Union's second largest park. It provides additional recreational opportunities for visitors to one of the most intensively used vacation and tourist regions in the country.

The Shatskiy park was the second to be created in the Ukraine. Its 82,500 ha lie in the northwestern part of the republic in Volyn province, near the Belorussian and Polish borders. First suggested in the 1970s, it is predominantly composed of wetland forests and

Figure 9.3   An early management proposal for Samarskaya Luka National Park

marshes in the upper watershed of the Pripyat River. Not being situated near any major cities, its functions seem to favor wetland protection first and tourism second (Nikolayevskiy, 1985).

In 1984 the Samarskaya Luka park was established along the middle Volga, in the Samara Bend area west of Kuybyshev. The middle Volga here enjoys its only brief undammed stretch, and steep bluffs on the inside of the bend provide spectacular vistas. A park has been contemplated here since the early seventies (Figure 9.3). Planning the area has been difficult, however, since it lies partly within the Volga-Ural oilfields, and over half its territory belongs to collective and state farms (Nikolayevskiy, 1985; Tezikova *et al.*, 1985).

Bayanaul'skiy is currently the only national park in the large Kazakh republic. Located in Pavlodar province, it was created in 1985 and extends over 45,500 ha. It takes in a remnant forested area, unusual rock formations, and several lakes, including Lake Dzhasybai, and receives about 40,000 visitors a year (Bugayev, 1985). The Kazakh Ministry of Forestry will develop it, like the other national parks, with both protected zones and zones for tourism.

The remaining national parks listed in Table 9.1 are all in the Russian

Figure 9.4   A portion of the Pribaikalskiy National Park along the west shore of Lake Baikal

republic. Mariy Chodra was created in 1985, the rest in 1986. Because of their recent establishment, little information is available about most of them. Mariy Chodra is located in the Mariy Autonomous Republic (ASSR), and takes in 36,600 ha. The Bashkir park is in the autonomous republic of the same name, and encompasses 79,800 ha in the mountains of the southern Urals. The Priel'brus'ye park extends over 100,000 ha along the north slope of the Caucasus Mountains in Kabardino-Balkarskiy ASSR. Its creation may have resulted in part from earlier calls for a "glacier national park" in the Caucasus (Tushinskiy, 1980).

Two parks were created in 1986 along the shores of Lake Baikal, finally fulfilling the promise of twenty years earlier. They are named Pribaikalskiy and Zabaikalskiy, referring to their respective locations on the near and far shores of the lake, as viewed from the west (Figure 9.4). The Pribaikalskiy park is located in Irkutsk province on 136,000 ha, and encompasses an elongated belt of shoreline between the town of Kultuk and the headwaters of the Lena River (Vorob'yev and Martynov, 1989). Zabaikal'skiy is situated in the Ust'-Barguzinskiy area of the Buryat ASSR, and is the largest national park in the Soviet Union at 209,000 ha. It also includes 370 sq. km of Lake Baikal itself. It is apparently adjacent to the Barguzin *zapovednik*, and the management of the two areas will be coordinated by the Siberian Division of the Academy of Sciences (*Sobr. post. prav. RSFSR*, 1986, No. 24).

References were encountered at the end of the 1980s to a possible new national park on the Baltic coast in Kaliningrad province. Its name has been given as Kurshskaya Kosa, and it may include the spit of land that extends almost 100 km along the Baltic Sea coast. In 1990, reference was also seen to a new Shorskogo national park of 455,000 ha in Kemerovo province in West Siberia.

The future of national parks in the USSR

The network of Soviet national parks is expected to continue to expand, and by the year 2000 there may be as many as thirty to forty located throughout the country. It seems probable that all fifteen Union republics will have established at least one national park by that time.

A number of suggestions for new parks have been advanced in recent years. These have included proposed parks in the Leningrad region, in the Tadzhik republic, along the Northern Donets river, a second one in Lithuania near Vil'nyus, as well as an "Obikhingou" national park in the glaciers of the Pamirs (Kotlyakov and Suprunenko, 1979). Nikolayevskiy in his volume suggests fourteen additional parks, including Lake Baikal. Past history, however, suggests that such recommendations are only modestly successful. Of the thirteen sites suggested in the 1967 work by Belousova, only about four have been established and in each case the locations or boundaries are different from what was originally recommended. As noted earlier, there have also been calls for the creation of historical national parks (Reimers, 1982).

An interesting approach for a new national park was advanced in 1988. This would in fact be an international park, including areas on both sides of the Bering Strait between Alaska and the Soviet Union's Chukchi Peninsula (*National Parks*, Nov./Dec. 1988). On the American side it would probably take in the present Bering Land Bridge National Preserve; there are numerous cultural sites on both sides of the Strait. Although only a concept in 1988, both sides found the idea of an international park worth pursuing.

Internally, a major need of Soviet national parks is more funds for park development. In this instance, such funding would be well rewarded, for it would be paid back in increased foreign tourism. This represents a real opportunity, particularly in the Baltic national parks, with their relatively easy access to western Europe. But after eighteen years of existence, none of them even have a visitors' center as yet.

Since the national parks are the creation of the Union republics, it may be considered inappropriate, or at least unnecessary, to create a nationwide national park management agency. However, it would seem desirable for the Russian Republic, with its large number of parks, to give consideration to the creation of a centralized park management authority. It is unfortunately probably the case that the present absence of such an agency in the Russian Republic reflects the relatively low priority given these parks in the budgetary process; as a result they are understaffed and face many potentially serious administrative problems (Kolbasov *et al.*, 1987). Yet there is much to recommend such a step: standardization of policies, reduction in certain overhead costs, better scientific cooperation, and more effective lobbying within Gosplan all come quickly to mind. The national (and perhaps international) prestige of the system might be enhanced, as well.

The rapid rate of creation of new national parks in the 1980s seems to indicate that the concept is currently a popular one both in the countryside and within the republic Councils of Ministers. It probably also reflects increased levels of both leisure and mobility within the Soviet population in recent decades. In all probability, before long these parks will also become the object of an increasing amount of visitation by foreign tourists to the Soviet Union.

10 Managing wildlife and endangered species

The challenge of wildlife conservation has taken on a greatly increased sense of urgency over the last two decades. Some of the reasons relate to a sense of lost esthetic values and others to the ethical aspects of extinction, but an even more compelling reason is now often advanced: self-preservation. The inhabitants of every country on the globe are today irreversibly dependent upon the genetic information contained in the world's resource bank of plants and animals. Although perhaps unaware of it, they utilize these biotic resources not only for foodstocks, but also as raw materials for clothing, medicine, and industry. Additionally, this genetic information is also widely used in the basic research being performed in all these areas, research upon which future generations will be increasingly dependent (Myers, 1983).

An important fraction of these genetic secrets, including both those that have been discovered and the larger fraction still unknown, reside within the flora and fauna of the Soviet Union. These flora and fauna are marvellously diverse (Knystautas, 1987). Natural scientists in the Soviet Union are aware of, and concerned about, the problem of conserving this wealth of genetic information. In this chapter, threats to the wildlife resources of the Soviet Union will be examined.

Causes of wildlife depletion

Collectively, the economic changes that occur as a nation industrializes produce the serious problem known as loss of habitat. Habitat reductions can be viewed as having three equally important components: loss of breeding grounds, loss of wintering areas, and loss of habitat critical for migration patterns.

The creation of preserved areas as a means to help conserve wildlife habitats has been reviewed in chapters 8 and 9. But the vast majority of the USSR lies outside these preserves. Here, the construction of new cities and industrial nodes, the expansion of agriculture, and the proliferation of mineral and timber extraction all decrease the percentage of the USSR that retains suitable natural habitat for wildlife. As a result, the Soviet Union, like many other nations, is facing serious depletions in its wildlife populations.

Agriculture poses some of the most significant threats to breeding habitats. In the Soviet Union, just since the 1950s, millions of hectares of land have been brought under cultivation in Siberia, Kazakhstan, Central Asia, and elsewhere, replacing the naturally occurring flora and fauna. In the steppe biome, little remains of the original vegetation cover, and animals that used this natural zone as their breeding habitat are now greatly reduced in numbers. The proliferation of domestic grazing animals has pushed many native species into smaller and more marginal habitats, further reducing their numbers. In some areas, so have recreational developments.

Industry has had a similar deleterious effect. Throughout the country mining and other industrial operations have steadily appeared, accompanied by scores of new urban centers, which together have transformed additional millions of hectares of former wildlife habitat. In addition to the land transformation, various forms of pollution typically follow as well, causing deterioration of such habitat as may still remain.

A good example of the loss of wintering habitat can be seen in the case of the red-breasted goose (*Branta ruficollis*), which breeds along the Arctic coast of the USSR (Figure 10.1). In between their tundra breeding seasons, as many as 25,000 of these geese wintered in the wetlands around the Araks and Kura rivers in Transcaucasia as recently as the early 1960s. But as the rivers became dammed and diverted in the interests of irrigated agriculture, by 1970, the wintering birds numbered only several hundred, and by the mid-1970s, only a few dozen (Vinokurov, 1986). Although the red-breasted goose has other wintering areas such as the Danube delta, few of these areas are well protected. The result has been that its total numbers have declined so sharply that it is now listed by the Soviet Union as a threatened species.

An unusual case of economic development adversely affecting wildlife migration routes occurred near the Arctic Siberian city of Norilsk. When major pipelines were built from distant natural gas fields into Norilsk, inadequate provisions were made for migratory reindeer to cross the lines. Confused, reindeer sometimes found themselves trapped between two parallel pipelines, with the result

Figure 10.1   Red-breasted goose (*Branta ruficollis*), a threatened species of the Soviet Arctic

that whole herds of them were funneled into downtown Norilsk. There, they became the accidental victims of automobiles and the intentional victims of poachers. Bureaucratic delays reportedly hindered finding a solution to this problem for years (Roslov, 1974).

Close behind loss of habitat as a cause of wildlife depletions is pollution of the environment. Effluents from industries and cities degrade numerous rivers, while waste products from agriculture and mineral extraction render additional large areas unsuitable for wildlife. Several reviews of Soviet water pollution have previously been prepared (see chapter 5), but the situation today remains serious. For example, the Amu-Darya River is presently home to three species of endangered fish, but is significantly polluted from agricultural run-off (Grachev, 1985). Although this pollution undoubtedly contributes to their endangered state, the main adverse factor has been that the biologically rich channels of the lower Amu-Darya and Syr-Darya rivers have been severely desiccated by irrigation water diversions (chapter 12). The harvest of muskrat skins in the delta of the Amu-Darya fell from 650,000 in 1960 to only 2,500 in the late 1980s (Micklin, 1988).

Economic development can have an adverse impact on fish and wildlife in a variety of ways. Over-extraction is one common problem; as one case in point a number of different commercial species in the Barents Sea have been heavily over-fished (Zelikman, 1989). Earlier, herring had been almost depleted by over-fishing in the Sea of Okhotsk (Kamentsev, 1984). The damming of the Volga has sharply reduced the catch of sturgeon in the Caspian basin, while discharges from the Cherepovets fertilizer plant killed millions of rubles worth of fish in local rivers in the early 1980s (Kozlov, 1984). Chapter 6 notes the 1983 fish kill in the Dnestr River caused by the collapse of a holding dam, as well as the death of wildlife in the southern Urals from an accidental release of high level radiation.

Some of the worst cases of environmental pollution throughout the world result from the use of pesticides. Chapter 6 noted that these intended aids to the economy have been increasingly employed in the Soviet Union and, as in other countries, their application has often had highly adverse effects on wildlife. Although the Soviet Union has conducted research on biological controls for over two decades, chemical pesticides continue to be heavily employed, and wildlife deaths from their use are still periodically reported.

These selected examples of the problems posed to wildlife by modern society can be multiplied many times over. The end result is that wildlife, to quote one Soviet observer, "is dwindling before our eyes" (Komarov, 1980). To date, no effective solution to the problem of the loss of habitat seems to have been found. Well over half of the Soviet Union has been transformed by agriculture, urbanization, forestry, and mining, and much of what has not been transformed lies in areas of relatively low biological productivity.

Inland fisheries

A similar litany of concerns about habitat loss and pollution is also applicable to the Soviet Union's inland fisheries. Their decline is easily quantified: prior to World War II, the catch from inland water bodies exceeded that from oceanic waters; today the ratio is ten to one in favor of the oceans. Catches of fish have decreased in the Sea of Azov and the Caspian by over 90 percent (Kamentsev, 1984). The Sea of Azov, in fact, was estimated in 1980 to be producing only about one-ninetieth of the postwar catch (Komarov, 1980, p. 38).

The causes of the decline in harvest from lakes, rivers, and estuaries includes such recurring themes as over-fishing (for example, at Lake Baikal), urban and industrial effluents (in numerous rivers), dams and reservoirs (Volga, Dnepr), increased salinity (Sea of Azov), and diversion of water for irrigation (Don, Syr-Darya, Amu-Darya, Kura, and other rivers). Chapter 2 also noted the adverse effect of increases in rainfall acidity on the health of lakes and reservoirs. These factors, singly or in combination, have greatly diminished fish stocks in a number of water bodies that historically were very productive.

The USSR has responded to these problems in a variety of ways, including the introduction of new species, constructing new fish breeding facilities and spawning areas, creating a vast network of small stock ponds, and occasionally instituting fishing moratoria where appropriate. It has even been suggested that specialized types of shellfish should be bred and introduced that would scavage and presumedly thrive in polluted waters (Narimanov, 1981). In response to the perceived seriousness of the problem, special resolutions on fisheries conservation and management were passed in 1974 and 1981 (Galeyeva and Kurok, 1986, pp. 228 and 371). Such actions have partly alleviated the problem, but the take of many desirable species (most notably sturgeon) continue to be well below former levels, and the Ministry of Fisheries acknowledged in 1984 that serious problems remain (Kamentsev, 1984).

The sturgeon, which is the source of caviar, represents an important case study. Caviar is extracted from three species of sturgeon, sevruga, osetr, and the giant beluga. The decline in the sturgeon catch was steady over several decades, dropping from 40,000 tons per year early this century, to 21,500 tons in 1956, to as little as 11,000 tons in the 1970s. Large hydroelectric dams on the Volga and the consequent loss of spawning grounds have been the main problems (Shipunov, 1987). Poaching, discussed below, has also been substantial, and the 100 to 400 ruble fines for illegally taking sturgeon are of debatable effectiveness (prison terms are possible and are used in severe cases).

With help, however, sturgeon in the Caspian basin are today reported to be making a strong comeback. At least a dozen new sturgeon farms have been built around the shores of the Caspian Sea, and from these hundreds of millions of new sturgeon fingerlings have been released into the Volga–Caspian system (Whitney, 1979). The various species of sturgeon are being cross-bred to produce larger and more stable harvests of caviar, and pre-war levels of sturgeon catches (20,000 to 30,000 tons) were reached during the 1980s.

Controlling hunting and poaching

The effective management of hunting and fishing by Soviet authorities remains an elusive goal. Poaching on a variety of species is as widespread today as in decades past. Even sanctioned forms of hunting and fishing are often inadequately controlled, and professional hunters, those who are licensed to take animals for their meat or hides, are often cited for over-hunting as well. The Soviet Union has enacted numerous laws designed to protect wildlife, but insufficient personnel, meager funding, and overly lenient enforcement causes them to be frequently ineffective (Baklanov, 1984; Pryde, 1986).

The decline in wildlife numbers would seem to suggest a need to tighten controls over sport hunting. Soviet sources indicate that hunting is not nearly as effectively managed as might be desired; Komarov in his chapter on wildlife is especially outspoken in this regard. He cites, as one example, a report that of 300 ducks banded on Lake Chany in West Siberia in 1976 not one survived to breed; all of them had been taken by hunters (1980, p. 76). *Izvestiya* reported that just between 1980 and 1984, 98 percent of the muskrat families and 95 percent of the beaver lodges in the Kiev Reservoir were eliminated by hunters (Baklanov, 1984).

The most common of the poaching problems appear to affect deer, elk, and sturgeon. The pervasiveness of the poaching problem is evidenced by the large number of articles that have appeared in the Soviet press on this subject over the past two decades. One writer noted that in a single year there were 943 poaching arrests in the republic of Byelorussia alone. Even more surprising in the Soviet context, in an eighteen-month period in the same republic 7,000 unregistered hunting guns were discovered (Shimanskiy, 1975). Another source reported that in 1983, 10,376 poachers received fines in just Yaroslavl and Vologda provinces alone (Kozlov, 1984). These sources also note that illegal hunting occurs even in nature reserves.

The seriousness of the poaching problem has been compounded (both in the Soviet Union and elsewhere) by the offenders including not just over-eager "sportsmen," but government officials as well (Komarov, 1980; Volkov, 1980). In addition, the killing of game wardens has been reported, with at least four such accounts appearing over a recent three-year period (*Nedelya*, Aug. 16, 1982; *Izvestiya*, Apr. 13, 1983 and June 3, 1984). In the 1983 account, the killer, who had shot the warden while being arrested, received only a fifteen-year sentence. In the Ukraine, wardens are not even allowed to wear sidearms (Baklanov, 1984).

Another case that illustrates well the dimensions of the problem occurred in Kazakhstan. A group of organized poachers were arrested for illegally killing 5,834 marmots, with the apparent complicity of officials of the Kazakh Administration for Hunting and Preserves. The poachers were arrested, but their sentences were so light that public outcry forced a retrial before the Kazakh supreme court. The retrial resulted in an increase in the prison term for the poachers from three to ten years; however, the officials involved received only reprimands (Prokhorov, 1983). Either stronger laws or stronger enforcement clearly seems to be needed.

In all, at least sixteen articles on the subject of poaching were translated in *Current Digest of the Soviet Press* from 1982 to 1984 alone. Such media attention, however, has not resolved the problem. Harsher sentences may help, but it would also appear that the Soviet education system must do a better job of persuading the public of the reasons for wildlife conservation. This reeducation on behalf of genetic preservation should include efforts directed at those who serve in the central governmental decision-making bodies, as well as the general public.

Soviet wildlife species of particular interest

Several species of wildlife in the Soviet Union have such an interesting history that they merit special mention. Some of them, like the European bison, involve longstanding conservation efforts that have been quite successful. Others have been more controversial, or may take more time. The following paragraphs review the status of a few of the more significant species.

The farthest north that tigers are found in the world is in the Soviet Union. Formerly, two sub-species of the Asiatic tiger existed within the USSR, *Panthera tigris virgata* and *P. t. altaica*. The latter is better known by its common name as the Siberian (or Amur) tiger, and today numbers about 200 in the Soviet Union and 600 worldwide. It is protected in the Sikhote-Alin *zapovednik* and in zoos around the world (Figure 10.2). Its presently increased numbers have reportedly led to livestock losses, as well as one tiger being found in 1986 in the middle of downtown Vladivostok. The other sub-species, called the Turanian tiger, was at one time fairly common in the valleys and deserts of Central Asia. As a result of hunting and economic development, its numbers rapidly declined in the twentieth century, and sometime between 1940 and 1960 it became extirpated in all portions of its former range (Borodin, 1985).

The European bison (*Bison bonasus*) represents one of the most determined of Soviet wildlife restoration efforts. This species is a close relative of the American bison but is less massive and is considered a separate species. Following World War I, the revolution, and the famine of the early 1920s, there remained within the USSR no female bisons young enough to breed. The re-establishment of the breeding stock was begun by borrowing a few females from Poland; for a more complete history of this effort the interested reader is referred to Pryde, 1972. Today, there are around 2,000 full-blooded European bison in the Soviet Union.

Probably the most controversial species found within the USSR is the European wolf (*Canis lupus*). Its problem is not that it is endangered; rather, like the American coyote, it is too successful a predator. As a result, within the USSR many consider it to be little more than an economic nuisance; "Peter and the Wolf" conveys the general image well. Prior to the Soviet Union's "environmental awakening" in the late 1960s, it was common practice to shoot wolves on sight, even within *zapovedniki* (Pryde, 1972). Since that time, numerous articles favoring preservation of the wolf have been

Figure 10.2 Siberian tiger, native to the southern portion of the Soviet Far East

published, and hunting has been restricted (Komarov, 1980). However, in the 1980s articles have again appeared that advocate reductions in wolf populations (Makeyev, 1982). In the USSR, the resourceful "gray bandit" may have a hard time improving its traditional negative image.

Wildlife species that migrate widely, such as many genera of fish, birds, and marine mammals, pose some of the most complex problems. As noted earlier, they require the protection of three habitats: breeding, migrating, and wintering. Cranes provide an instructive example. Four species of cranes out of the seven found in the USSR either are or have been on the endangered list. One of these, the Siberian white crane (*Grus leucogeranus*), nests only in the Soviet Union, in northern Yakutia (Flint, 1984). As yet, it has no specific nature reserve set up within its breeding territory. Probably less than 400 Siberian cranes now exist in the world. The other three threatened crane species nest over a broader area in the far east, and are protected at the Zha Long and Wucheng Nature Reserves in China, and elsewhere. All, however, are inadequately protected in their migration and wintering habitats (Stewart, 1987).

The saiga (*Saiga tatatica*) is a species of steppe antelope with a seemingly over-sized nose. Although not presently found in the red books, it is a popular target of hunters and has long been of widespread conservation concern. Komarov, writing in the late 1970s, noted that in the sparsely populated Kalmyk ASSR, saiga have been killed by the hundreds on fences and in irrigation canals (Komarov, p. 80). Almost a decade later, a special meeting of the USSR Scientific Council was called to consider the continuing plight of the saiga in the face of both economic development and poaching (Pralnikov, 1985).

A final example is the snow leopard (*Panthera uncia*). Although enjoying a wide range throughout the Pamir, Himalaya, and Sayan mountains, it is easily displaced and is now listed as rare in the Soviet Red Book. Its elusiveness can be seen in the wide range of figures for its current USSR population – from 300 to 1,000 (Braden, 1982). It is protected in a number of reserves in the Soviet Union, and is also the subject of a significant international preservation effort. The city of Alma-Ata has hosted international snow leopard conferences, most recently in 1989.

Managing endangered fauna

Species extinction is the ultimate consequence of the above assaults on wildlife populations, and constitutes a rapidly accelerating problem in all parts of the world. As natural habitats give way to the various encroachments of civilization, numerous types of fauna are inevitably pushed to the edge of elimination. Species that once resided within the present area of the USSR which are known to have become extinct over the past two centuries include the Stellar's sea cow, spectacled cormorant, Turanian tiger, and probably the crested shelduck. Soviet scientists began expressing concern over potentially endangered species as early as 1930 (Weiner, 1988, p. 122). Today, the protection of threatened species has become a matter of considerable concern in the Soviet Union, as it has elsewhere.

An important part of the USSR's wildlife conservation program has been the publication of an official "Red Book," in which are described those species of native flora and fauna that are in decline or are threatened with extinction. The earliest prototype Red Book in the USSR was published in 1928 (Weiner). The first official edition of a Soviet Red Book (*Krasnaya kniga SSSR*) appeared in 1974, and has since undergone two revisions (Borodin 1978, 1985). In addition, most of the fifteen Union republics have produced their own local Red

Table 10.1 *Rare, declining, and endangered fauna in the USSR*

CLASS	IUCN Category					
	I	II	III	IV	V	n=
Mammals	23	18	40	10	3	94
Birds	21	25	16	14	4	80
Amphibians	1	6	2	0	0	9
Reptiles	7	7	16	6	1	37
Fish	7	0	1	1	0	9
Insects	9	75	99	19	0	202
Crustaceans	0	0	2	0	0	2
Bivalves	1	1	13	0	0	15
Gastropods	0	0	4	0	0	4
Worms	1	1	9	0	0	11
Total	70	133	202	50	8	463

Sources: Borodin, *Krasnaya kniga SSSR* (Part 1), 1985

Books. The 1985 edition of the Soviet Red Book appears in two volumes, the first for fauna and the second for flora. The fauna volume lists a total of 463 species (or sub-species) of wildlife, primarily vertebrates (Table 10.1). Preparation of future Red Books has recently been made the responsibility of the new State Committee for Environmental Protection (Goskompriroda).

For purposes of classifying species of flora and fauna that are endangered or otherwise low in numbers, the USSR Red Book utilizes the status categories created by the International Union for the Conservation of Nature and Natural Resources (IUCN). These categories are defined in Appendix 10.1. Of the 463 species or sub-species in the Red Book, 70 have been given the most critical status of "endangered" (i.e., IUCN category I).

The fauna that are classified as endangered in the Soviet Union's Red Book include 23 species or sub-species of mammals, 21 species of birds, 7 reptiles, 7 fish, 9 insects, and 1 each amphibian, bivalve, and worm (lower [more primitive] classes of faunal life are not covered). The complete list is given in Appendices 10.2, 10.3, and 10.4.

A composite distribution map, illustrating where all seventy of the Soviet Union's endangered species and sub-species of fauna are to be found, has recently been prepared (Pryde, 1987). It reveals a very uneven spatial pattern (Figure 10.3). In general, the southern portion

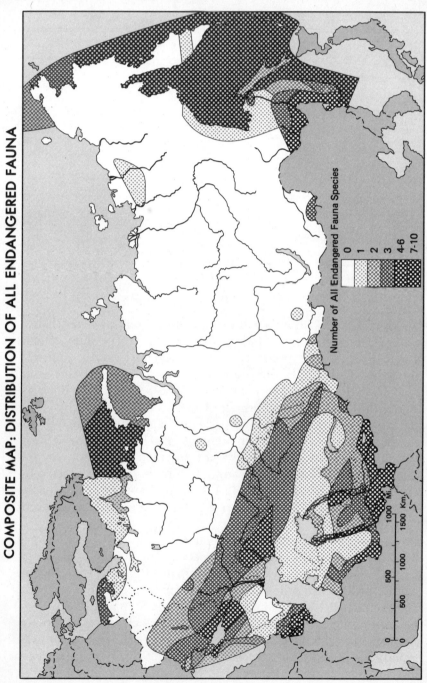

COMPOSITE MAP: DISTRIBUTION OF ALL ENDANGERED FAUNA

Number of All Endangered Fauna Species

0
1
2
3
4-6
7-10

Figure 10.3   Distribution of endangered species in the USSR

of the Soviet Far East, the major seas on the western and eastern extremities of the country, the Transcaucasus, all portions of Central Asia, and a few other discrete areas account for the vast majority of the seventy species.

The three Baltic republics, Belorussia, and the northern portions of the Russian republic are all portions of the USSR containing relatively few species listed as endangered in the Red Book. The heavily industrialized areas of the Central (Moscow) region, the Urals, and West Siberia show no endangered species at all, most likely because large areas of similar habitat surround them. Nor does Lake Baikal contain any, despite its long-standing water quality controversies (chapter 5).

In terms of natural biogeographic zones, several stand out quite clearly as having a high concentration of endangered species. These are for the most part located along the southern perimeter of the country, and include the steppe and desert steppe, Crimean and Caucasus mountains, desert, Central Asian mountains, and Far East mixed forest zones. Also ranking high are such maritime areas as the Barents, Baltic, Okhotsk, Japan, and Bering seas, due to the presence there of several threatened whale and other marine mammal species. For ease of reference, the totals of endangered species by geographical region are presented in Table 10.2.

The vast tundra and taiga zones are virtually free of endangered species, with only localized exceptions appearing on the maps. This, however, was not quite so true for earlier editions of the Red Book, which included as endangered such northern species as the gyrfalcon and two species of tundra-nesting geese. These were dropped from Category I in 1985, apparently reflecting an improved status.

Effective public education regarding endangered species is critical to the success of the effort. The concept of genetic preservation has appeared in Soviet scientific works since around 1980, but only much more recently in books for the general public. The popular science-nature magazine *Priroda* frequently carries articles spotlighting some particular species found in the Red Books. Special books have also begun to appear that endeavor to educate the Soviet public about endangered species. Zoos and natural history museums help educate the public on wildlife conservation issues, as well. Nevertheless, given the extent of the poaching and similar problems, all of these efforts need to be expanded.

To assist in protecting the species found in the Red Books, both effective legislation and sufficient levels of funding are needed. At

Table 10.2 *Distribution of endangered Soviet fauna by biogeographical region*

Biogeographical region	23 Mammal species[a]	21 Bird species[a]	26 Other species[a]	Totals by region
1 Tundra and forest-tundra		1	1	2
2 Taiga	1		1	2
3 European mixed forest			2	2
4 European and West Siberia forested steppe		1	4	5
5 European steppe and desert steppe			6	6
6 Crimean and Caucasus mountains and steppe	2	4	10	16
7 Kazakh steppe		2	3	5
8 Central Asian deserts	1	5	5	11
9 Central Asian and Kazakh mountains	5	4		9
10 Mountains of southern West and East Siberia	1	2		3
11 Mongolian steppe	1	2		3
12 Amur-Primorskiy Region	4	8	1	13
13 Kamchatka – Kuriles		2		2
14 Sakhalin Island		2		2
15 Baltic Sea	3			3
16 European Arctic Pelagic	4			4
17 Asiatic Pelagic	5	1		6
18 Black Sea	1		1	2
19 Central Asian rivers			4	4
20 Caspian and Aral Seas			b	0

Notes: [a] Columns do not total to number of species listed at column head, as several species occur in more than one biome. In some cases, a species occurs over most or all of an indicated region, but in other cases may be found within only a small portion of it
[b] One species, now believed extirpated, used to occur in the Aral Sea; none occur in the Caspian

Source: Pryde, 1987, Figures 1 through 3

present, the Soviet Union has no single law containing the mandatory conservation provisions (including mitigation measures) of the United State's Endangered Species Act. Such mitigation is implied, but not explicitly required, in certain key Soviet enactments, such as the USSR's wildlife protection law of 1980 (discussed in the next section). Although existing laws provide a certain amount of preservation assistance for threatened species in the Soviet Union, it is probable ythat a stringent Soviet version of the Endangered Species Act might provide more effective assistance. Such provisions will hopefully be included in the new Law on Environmental Protection, which was being drafted in 1989.

Conservation laws and agreements

The Soviet Union has enacted a significant body of laws to advance the cause of wildlife preservation. The 1977 Soviet constitution itself, in Article 18, specifically mandates the "scientific, rational use of the . . . plant and animal kingdoms . . . " The wording in this section is necessarily general, and it can certainly be questioned as to how much effect it has had, but at least the thought is there. Many people have suggested that the right to a healthy environment might be a worthwhile addition to the Bill of Rights in the United States constitution.

The first comprehensive environmental legislation appeared in the Soviet Union during the 1960s. At that time, all fifteen of the Soviet Union's constituent republics passed generalized conservation laws to guide their own natural resource management activities. Sections relating to wildlife conservation (often with a focus on commercially useful animals) were usually included. In these earlier acts, the wording was at times rather utilitarian in tone. For example, the right of wildlife to exist for its own sake appeared to depend on "good behaviour," for the 1960 Russian republic conservation law states that "it is forbidden to destroy noncommercial wild animals *if they do not harm the economy or public health*" (emphasis added). This unfortunate attitude stems from classical Marxist economics, which holds that resources have value only in proportion to their economic utility.

In 1980, for the first time, a national statute was enacted that governed wildlife management practices for the whole of the USSR. Entitled the "Law of the U.S.S.R. on the Protection and Utilization of the Animal World," it was translated in full in the August 20, 1980 issue of *Current Digest of the Soviet Press*. The "Law . . . " is comprised of thirty-nine articles which establish nationwide guidelines, policies, and jurisdictional authority over the regulation of hunting, fishing, and wildlife management and preservation. The section outlining basic requirements for wildlife conservation (Article 8) is translated as Appendix 10.5. A welcome change from the utilitarian attitudes noted in the previous paragraph can be seen in its wording.

The "Law . . . " contains the most explicit section to date dealing with endangered species. Article 26 of this act states, among other provisions, that actions which would cause the destruction of rare or endangered species of animals, or of their habitats, are not permitted (Appendix 10.6). Implementation of this section in the context of specific construction projects is, of course, another matter. More generalized wording to the same effect, stating that "rare and endangered species of animals are also subject to protection against destruction and extinction," had appeared twenty years earlier in the 1960 Russian republic law (Pryde, 1986, Appendix 3).

The implementing legislation to carry out all of the policies in the national law is enacted at the Union republic level. The regulation of hunting, habitat protection, establishment of preserves, etc., for the most part are Union republic decisions (although the more important of them may require the consent of the state planning agency, Gosplan, or the Council of Ministers). An important part of such implementing legislation is enforcement. The earlier section on poaching gave indications that enforcement is sometimes lax, and the effectiveness of enforcement is often questioned in Soviet publications. One highly regarded Soviet environmental law expert has cited the Ukraine specifically as a case in point (Kolbasov, 1983, p. 117f).

Soviet wildlife conservation efforts extend well beyond their own borders. Since the 1960s, the USSR has regularly participated in international conferences on environmental issues, and has signed numerous agreements relating to over-harvested species (mainly oceanic fish), and to endangered and threatened species such as the polar bear (chapter 15). The USSR has long played an active role in the work of the International Union for Conservation of Nature and Natural Resources (IUCN), the UN's Man and the Biosphere program, as well as the International Council for Bird Protection, the International Crane Foundation, and many similar organizations (Knystautas, 1987).

The US–USSR environmental exchange agreement signed in 1972 included many provisions relating to wildlife preservation. The two countries also have bilateral agreements that govern fishing and crab harvesting in northern Pacific waters, as well as others to protect migrating birds of mutual concern; similar agreements exist with Japan (Galeyeva and Kurok, 1986, pp. 229 and 328), and with many other countries. The Soviet Union has also signed the CITES (Convention on International Trade in Endangered Species) agreement, and has ratified the Ramsar Convention which protects wetlands used by migratory waterfowl (see chapter 11).

Another legislative creation that protects wildlife is the Soviet Union's network of nature reserves, or *zapovedniki*, which were described in chapter 8. Among the species that have been helped by these reserves are the beaver, European bison, sable, saiga, otter, polar bear, and Siberian tiger (see Table 8.4). On a few of the *zapovedniki*, however, the scientific research function is so dominant, or the total area of the preserve is sufficiently limited, that a comprehensive species protection program is not feasible, and in some cases resident species that were formerly common have sharply declined (Yeliseyeva, 1976). The network of *zapovedniki* will continue to play an important role in protecting wildlife habitats, but as chapter 8 noted, they cover too small a portion of the country's total area (about 0.5 percent) to represent a sufficient response by themselves. Only if the *zapovedniki* network were to be expanded to at least twice its present size, *with the new preserves located exclusively in the most critical wildlife areas*, would it approach being a viable response to the wildlife habitat problem.

Criticisms of Soviet whaling practices heard in the 1970s and early 1980s have largely been eliminated by decisions made in the late 1980s to comply with the International Whaling Commission's restrictions on whaling activities (Ziegler, 1987, p. 146). However, in 1988 the Environmental Defense Fund reported that the USSR acknowledged taking 5,000 seals from Antarctic waters; the reasons for this large harvest were not specified.

Soviet wildlife: a summary

Wildlife conservation in the Soviet Union is characterized by both noteworthy success, as well as some serious problems. Many of the successes involve endangered species, which presently are receiving increased emphasis following the publication of the Soviet Red Books. The problem areas include loss of habitat (especially steppe and wetlands), protecting endangered species, reducing poaching, mitigating the effects of economic development, and, closely related to all the foregoing, implementing and enforcing laws. Soviet planners are not oblivious to the need for wildlife preservation, but conservation progress is generally hard-pressed to keep up with economic progress. As a result, additional endangered species will appear in the future.

In the past, economic planning usually gave little attention to wild-life considerations. Soviet zoologists, although their voices were heard, carried relatively little weight in the halls of the state planning agencies. There are signs this may be changing.

Today, where conservation difficulties exist, it is not because the scientific community is inattentive, nor because they have neglected to alert the appropriate ministries and officials to these problems. Rather, the economic agencies, burdened by the usual considerations of budgets and priorities, know that fulfillment of the economic plan, not wildlife preservation, will be the basis upon which their perform-ance will be judged.

This chapter has outlined some of the major Soviet laws enacted on behalf of wildlife conservation. Questions can be raised, however, as to the efficacy of enforcement of these laws, and the adequacy of budgets for enforcement personnel. Historically, the funding for both pollution abatement facilities and wildlife management activities has been chronically deficient. A higher level of public education as to the reasons and necessity for genetic conservation is also needed. The forthcoming Law on Environmental Protection may provide help in these areas.

A lingering problem from the past is that a utilitarian attitude towards nature was nurtured by decades of textbooks which included passages lauding the USSR's "inexhaustible natural resources" (see Pryde, 1972, p. 4). Recent textbooks generally treat natural resources in a more realistic manner (e.g., Astanin and Blagosklonov, 1983). However, almost all present adult Soviet citizens (i.e., those now in policy-making positions) grew up learning the old misleading maxims.

The USSR remains a nation rich in biological diversity. Yet, despite having taken many significant steps on behalf of biotic preservation, the relentless striving for economic development continues to keep wildlife in a compromisable and precarious position. For the future, more determined efforts, more education, more nature reserves, more enforcement, and more funding in the interests of wildlife pre-servation will need to be brought into play. Only in this way can the sixth of the earth's land resources which have been entrusted to the Soviet Union maintain their biological integrity and spectacular diversity.

Appendix 10.1
IUCN status categories

The IUCN system for classifying potentially threatened species employs seven categories. The five primary IUCN categories, as adapted for use in the Soviet Red Books, are defined as follows.

Category I (IUCN = E). Endangered. Species currently found under the threat of extinction; special measures are needed to assist their survival.

Category II (IUCN = V). Vulnerable. Species whose numbers are low, or are declining at a significant rate, and which may become threatened with extinction in the near future.

Category III (IUCN = R). Rare. Species that are not now threatened with extinction, but whose small numbers or limited territories could jeopardize them should sudden changes occur in their habitats.

Category IV (IUCN = I). Indeterminate. Species whose low numbers cause them to be of concern, but whose biology is inadequately studied to allow them to be placed into one of the first three categories.

Category V (IUCN = O). Out of danger. Re-established species that are increasing in numbers and are not presently of concern, but that still need protection and supervision.

The other two IUCN categories, Extinct (Ex) and Insufficiently known (K), are not employed by the Soviet Red Books.

The above five basic categories are used in the 1985 USSR Red Books, though they were not in the earlier editions. Table 1 indicates the number of listed species that have been placed into each of these five categories.

Appendix 10.2
Endangered species of mammals in the USSR (listed in order shown in 1985 USSR Red Book)

Asiatic river beaver (Aziatskiy rechnoy bobr)
(*Castor fiber pohlei*)
Red wolf (Krasnyy volk)
(*Cuon alpinus*)
Anatolian leopard (Peredneaziatskiy leopard)
(*Panthera pardus ciscaucasica*)
Amur leopard (Amurskiy leopard)
(*Panthera pardus orientulis*)
Turanian tiger (Turanskiy tigr)
(*Panthera tigris virgata*) – P. t. v. is now probably extinct
Cheetah (Gepard)
(*Acinonyx jubatus*) – possibly extirpated within USSR
Harbor seal – Baltic population only (Obyknovennyy tyulen')
(*Phoca vitulina vitulina*)
Baltic gray seal (Baltiyskiy seryy tyulen')
(*Halichoerus grypus macrorhynchus*)
Mediterranean monk seal (Belobryukhiy tyulen'; tyulen'-monakh)
(*Monachus monachus*)
Bottlenose whale (Vysokolobyy butylkonos)
(*Hyperoodon ampullatus*)
Gray whale – Sea of Okhotsk population only (Seryy kit)
(*Eschrichtius robostus*)
Bowhead whale (Grenlandskiy kit)
(*Balaena mysticetus*) Barents and Okhotsk populations only
Right whale (Yuzhnyy kit; Yaponskiy kit)
(*Eubalaena glacialis*)
Humpback whale (Severnyy gorbach)
(*Megaptera novaeangliae*)
Blue whale (Siniy kit)
(*Balaenoptera musculus*)
Spotted deer (Pyatnistyy olen')
(*Cervus nippon hortulorum*)

Mongolian gazelle (Dzeren)
(*Procapra gutturosa*)
Goral (Goral)
(*Naemorhedus caudatus*)
Markhor (Vintorogiy kozel; Markhur)
(*Capra falconeri*)
Altai mountain sheep (Altayskiy gornyy baran)
(*Ovis ammon ammon*)
Transcaucasus mountain sheep (Zakavkazskiy gornyy baran)
(*Ovis ammon gmelini*)
Kyzyl-kum mountain sheep (Kyzylkumskiy gornyy baran)
(*Ovis ammon severtzovi*)
Bukhar mountain sheep (Bukharskiy gornyy baran)
(*Ovis ammon bocharensis*)
Source: *Krasnaya kniga SSSR*, 1985

Appendix 10.3
Endangered species of birds in the USSR (listed in order shown in 1985 USSR Red Book)

Short-tailed albatross (Belospinnyy al'batros)
(*Diomedea albatrus*)
Japanese crested ibis (Krasnonogiy ibis)
(*Nipponia nippon*)
Far-eastern stork (Dal'nevostochnyy aist)
(*Ciconia boyciana*)
Swan goose (Sukhonos)
(*Anser cygnoides*)
Crested shelduck (Khokhlataya peganka) (may be extinct)
(*Tadorna cristata*)
Marbled teal (Mramornyy chirok)
(*Anas angustirostris*)
Pallas's sea-eagle (Orlan-dolgokhvost)
(*Haliaeetus leucoryphus*)
Lammergeier (Borodach)
(*Gypaetus barbatus*)

Lanner falcon (Sredizemnomorskiy sokol)
(*Falco biarmicus*) – possibly extirpated within the USSR
Barbary falcon (Shakhin; Pustynnyy sokol)
(*Falco pelegrinoides*)
Japanese crane (Yaponskiy zhuravl')
(*Grus japonensis*)
Siberian crane (Sterkh; Beliy zhuravl')
(*Grus leucogeranus*)
Houbara bustard (Dzhek;. Vikhlyay; Drofa-krasotka)
(*Chlamydotis undulata*)
Ibisbill (Serpoklyuv)
(*Ibidorhyncha struthersii*)
Spotted greenshank (Okhotskiy ulit)
(*Tringa guttifer*)
Slender-billed curlew (Tonkoklyuvyy kronshnep)
(*Numenius tenuirostris*)
Relict gull (Reliktovaya chayka)
(*Larus relictus*)
Blakiston's fish-owl (Rybnyy filin)
(*Ketupa blakistoni*)
Scaly-bellied green woodpecker (Cheshuychatyy dyatel)
(*Picus squamatus*)
Reed parrotbill (Trostnikovaya sutora)
(*Paradoxornis heudi*)
Jankowski's bunting (Ovsyanka Yankovskogo)
(*Emberiza jankowskii*)

Source: *Krasnaya kniga SSSR*, 1985

Appendix 10.4
Endangered species of other major classes of Soviet fauna[a]

Siriyskaya chesnochnitsa (grey spadefoot toad) (*Pelobates syriacus*)
Sredizemnomorskaya cherepakha (Mediterranean turtle) (*Testudo graeca*)

[a] Some of these species may have no standardized English name; in such cases only a roughly literal English translation or transliteration has been given, based on either the scientific or Russian name, whichever is more helpful.

Ruinnaya agama (ruinnaya agama lizard) (*Agama ruderata*)

Zakavkazskaya takyrnaya kruglogolovka (Transcaucasus takyr toad-headed lizard) (*Phrynocephalus helioscopus*)

Pyatnistaya kruglogolovka (spotted toad-headed lizard) (*Phrynocephalus maculatus*)

Maloaziatskaya yashcheritsa (Asia minor lizard) (*Lacerta parva*)

Bol'sheglazyy poloz (big-eyed snake) (*Ptyas mucosus*)

Leopardovyy poloz (leopard snake) (*Elaphe situla*)

Atlanticheskiy osetr (Atlantic sturgeon) (*Acipenser sturio*)

Bol'shoy amudar'inskiy lzhelopatonos (great Amu-Darya shovelfish) (*Pseudoscaphirhynchus kaufmanni*)

Malyy amudar'inskiy lzhelopatonos (little Amu-Darya shovelfish) (*Pseudoscaphirhynchus hermanni*)

Syrdar'inskiy lzhelopatonos (Syr-Darya shovelfish) (*Pseudoscaphirhynchus fedtschenkoi*)

Aral'skiy losos' (Aral salmon) (*Salmo trutta aralensis*)

Sevanskaya forel' (Sevan trout) (*Salmo ischchan*)

Volkhovskiy sig (Volkhov River whitefish) (*Coregonus lavaretus baeri*)

Yapiks gigantskiy (giant yapiks) (*Heterojapyx dux* – Order *Diplura*)

Tolstun stepnoy (steppe tolstun beetle) (*Bradyporus multituberculatus*)

Rozaliya izumrudnaya (emerald rosalia beetle) (*Rosalia coelestis*)

Morimus temnyy (drab morimus beetle) (*Morimus funereus*)

Sovka shpornikovaya (delphinium cut-worm moth) (*Chariclea delphinii*)

Zor'ka zegris (zegris "dawn" butterfly) (*Zegris eupheme*)

Shashechnitsa pustynnaya ferganskaya (Fergana desert checkerspot) (*Melitaea acreina*)

Golubyanka stepnaya ugol'naya (coal steppe "blue" butterfly) (*Neolycaena rhymnus*)

Pestryanka leta (mottled leta moth) (*Zygaena laeta*)

Zhemchuzhnitsa yevropeyskaya (European pearl-oyster) (*Margaritifera margaritifera*)

Eyzeniya gordeyeva (gordeyev eisenia worm) (*Eisenia gordejeffi*)

Source: *Krasnaya kniga SSSR*, 1985

Appendix 10.5
Excerpt from "Law of the USSR on the Protection and Utilization of the Animal World"

Art. 8. Basic requirements for the protection and utilization of the animal world

During the planning and implementation of measures that may have an impact on the habitat of animals and the condition of the animal world, observance of the following basic requirements is to be ensured:

the preservation of the species diversity of animals in a state of natural freedom;

the protection of the habitats of animals, conditions for their reproduction, and their migration routes;

the preservation of the integrity of natural communities of animals;

the scientifically substantiated, rational utilization and reproduction of the animal world; and

the regulation of the number of animals, with a view to protecting the health of the population and preventing damage to the national economy.

Source: "Law of the USSR . . . ," 1980

Appendix 10.6
Excerpt from "Law of the USSR on the Protection and Utilization of the Animal World"

Art. 26 The protection of rare and endangered species of animals

Rare and endangered species of animals are registered in the books listing rare and endangered species of plants and animals – the USSR Red Book and the Union-republic Red Books. Regulations governing the USSR Red Book are confirmed according to a procedure determined by the USSR Council of Ministers, and regulations governing the Union-republic Red Books are confirmed according to procedures determined by Union-republic legislation.

Actions that may lead to the death of rare and endangered species of animals, decreases in their numbers, or the disruption of their habitats are not permitted.

With a view to the preservation of rare and endangered species of animals whose reproduction in natural conditions is impossible, the specially empowered state agencies for the protection of the animal world and the regulation of its utilization are obliged to take steps to create the necessary conditions for the breeding of these species of animals.

The taking of rare and endangered species of animals for breeding in specially created conditions and subsequent release into the wild, as well as for scientific research and other purposes, is permitted by special authorization, issued by the specially empowered state agencies for the protection of the animal world and the regulation of its utilization.

Source: "Zakon SSSR ob okhrane i ispol'zovanii zhivotnogo mira," *Pravda*, June 28, 1980, pp. 1 and 3, as translated in *Current Digest of the Soviet Press*, vol. 32, no. 29 (Aug. 20, 1980), p. 13

11   Protecting the land

If there is any natural resource that the Soviet Union has in abundance, it is land. Unfortunately, people are often unconcerned about conserving anything that is perceived to exist in unlimited amounts. Yet land, whether we are talking about small tracts or entire extensive biomes, can be remarkably fragile and easily degraded. All societies, but particularly those to whom the greatest riches have been entrusted, have an obligation to exercise great care in the use of the land that sustains them.

Land resources may be classified into two main categories: renewable and non-renewable. The former include vegetation, wildlife, and most fresh water; the latter include mineral resources, and, in a human time scale, soil. The renewable resources have largely been reviewed in previous chapters; this chapter will look particularly at the conservation of soil in the context of agricultural and mining activities, as well as at some multiple-use aspects of managing the Soviet Union's extensive but vulnerable land resource that have not been previously discussed. In 1990 the USSR revised its basic land use statute ("Principles . . .," 1990); for the text of the earlier version see Pryde, 1972.

It is beyond the scope of this chapter to examine in detail the complexities of Soviet land use planning. The interested reader can find this topic discussed in the cited article by Denis Shaw (1986) and the volume by Pallot and Shaw (1981). This chapter will focus on some specific problems and management approaches that are of significance in the context of Soviet land use management and conservation.

The quantity and nature of the Soviet land resource

The total area of the Soviet Union is 22.27 million sq. km (8.6 million sq. miles), giving it a land resource two and one-half times that of the United States and almost one hundred times that of the United

Table 11.1   *The Soviet land base, by management categories*

Category	Million	ha	
All land	2,227.6		
including land managed by			
Agricultural users		1,049.6	
of which, agricultural land			(605)
non-agricultural land			(445)
Forestry users and State Land Reserve		1,107.0	
Other land users (urban, etc.)		71.0	

Source: Narodnoye khozyaystvo SSSR v 1987 g., 1988 p. 183.

Table 11.2   *Land area by natural zones*

Natural zone	Million ha
Arctic tundra	180
Taiga	755
Forested steppe	64
Steppe	257
Desert-steppe	130
Desert	173
Mountain areas	652
(Other/unclassified)	16
Total	2,227

Source: Adapted from Rozov, 1971, ch. 8, p. 108.

Kingdom (Table 11.1). One of the best indicators of the vast size of the Soviet Union is that its land mass extends over eleven time zones.

However, due to the generally northern location of the country, much of that land resource is rather limited in terms of its possible uses and potential habitability. Aridity and mountainous terrain are two other widespread constraints. Some of the largest natural zones found in the USSR, such as the Arctic tundra and the Central Asian deserts, not only have limited economic possibilities but also possess fragile and easily destroyed ecosystems (Table 11.2). As chapter 7 pointed out, even the largest natural zone, the vast forest belt or *taiga*, has important limitations on the extent and nature of its use, and has suffered from organizational mismanagement.

On the other hand, some of the most threatened ecosystems in the

Soviet Union are relatively small, and have already undergone a high degree of adverse transformation. Such ecosystems include the narrow but very important vegetated corridors along rivers (particularly those found in arid areas), as well as the rivers themselves. They also include coastlines, both tidal and freshwater wetlands, and isolated areas of rare or unique vegetation.

In the Soviet Union all land belongs to the state (at least as of 1989) and therefore both decisions on the use of the land and responsibility for its conservation are also functions of the state. These responsibilities are generally delegated to ministries that are supervised by either national or republic-level governmental bodies, such as those in charge of agriculture, timber harvesting, or mining.

The long-standing problem has been that both the implementing ministries and the overseeing governmental bodies have a built-in conflict of interest in being responsible for both meeting economic targets and protecting the environment. The agricultural sector, for example, has pursued policies that have sacrificed soil fertility in order to increase crop productivity (*Soviet Geography*, 1989, pp. 759ff). Over the country as a whole, 13 million ha of land suffer from soil erosion, 12 million ha have been lost to construction and roads, and 3 million to poor irrigation systems, 10 million have been flooded for reservoirs, and 2 million have been destroyed by open pit mining (Yablokov, 1988). As has been previously noted, the economic mandate has in general enjoyed dominance over conservation goals. As a result, many land use management problems, such as erosion and reclamation, have evidenced a rather slow pace of improvement.

The fight against erosion and deflation

The problems of soil erosion caused by the action of water and wind (the latter referred to as *deflation*) have been periodically examined in the Soviet context (e.g. Pryde, 1972; Stebelsky, 1987; Vasil'yev *et al*, 1988; Sazhin, 1988). Such studies have pointed out that the Soviet Union has been concerned about its serious soil erosion problem for a number of decades. Unfortunately, over that time period the problem has not been solved, although some progress has been made. It has also taken on some new dimensions. For example, the deflation problem now includes "salt storms" around the drying Aral Sea bed, a phenomenon that will be discussed in chapter 12.

Soil erosion not only represents an obvious loss of an important natural resource, but also severely affects farming efforts. Soviet agriculture, which has difficulty meeting the nation's food needs in even the best of years, cannot tolerate the added burden of continuing topsoil losses.

The Soviet Union is unfortunately highly vulnerable to soil erosion. Much of its arable land is either on sloping terrain which is susceptible to water erosion, or is in semi-arid regions that are subject to recurring drought and deflation. These concerns are exacerbated by the extensive occurrence of fine-grained, wind-deposited soils, called *loess*, over much of the fertile steppe and forested-steppe regions of the country. Past farming practices, going back into the nineteenth century, were often inadequate to protect these highly erodable topsoils, and losses in the pre-Soviet period were great (Stebelsky, 1987). In the twentieth century, sub-optimal agricultural practices (plowing marginal lands, soil salinization in arid areas, Khrushchev's emphasis on growing corn, improper use of fertilizers, downslope plowing patterns, use of excessively heavy machinery) have also often been conducive to topsoil degradation and accelerated erosion.

As one result of the above, there have been created extensive networks of water-eroded gullies and ravines throughout much of European Russia and the Ukraine. Recent studies have shown the potential for this kind of erosion to be widespread even north of the main agricultural zone, that is, even at the latitude of Moscow (Makhaveyev *et al.*, 1984). The water erosion problem has been attacked by a number of measures, primarily the widespread planting of soil-stabilizing vegetation so as to prevent the further spread of such gullying. In some areas these efforts have produced the desired results, but in other areas the problem remains acute, with the extent of the gullies slowly but steadily increasing.

In the more arid parts of the country, such as southern Ukraine, the southern Russian lowland (the area from the Sea of Azov to the lower Volga River), and the northern Kazakhstan to southern West Siberia region, the major problem has been wind-induced erosion, or deflation (Figure 11.1). Vast areas of very good soils have been adversely affected in the two prime dust storm regions, which are the south Russian lowland (900,000 sq. km) and northern Kazakhstan to southern West Siberia (700,000 sq. km). The dust storms in the West Siberia–Kazakhstan region tend to be shorter in duration than those in the south Russian lowland, and in the latter region there is also much more variation in the annual number of dust storms (Sazhin, 1988).

Such dust storms represent a major threat to Soviet agriculture. The largest of them, such as the ones in 1928 and 1960, have devastated

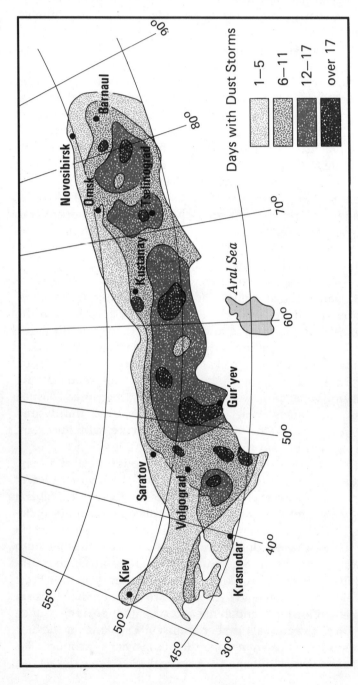

Figure 11.1 Frequency of dust storms (average annual days) in the Soviet steppe region
Source: Adapted from Sazhin, 1988, p. 936

vast areas. For example, the 1960 storm not only produced deflation over much of the southern Ukraine and North Caucasus, but also dropped dust over large portions of Eastern Europe, including parts of Poland, Hungary, Czechoslovakia, and Romania (Pryde, 1972, p. 40). In the south Russian lowland, severe dust storms have occurred in the postwar period in 1953, 1960, 1965, 1969, 1974, and 1984, with deflation episodes of lesser intensity being experienced in several other years. Some areas in the south Russian lowland have in excess of twenty days of dust storms annually, with from 2 to 7 cm of topsoil being removed in the worst cases (Vasil'yev *et al.*, 1988; Sazhin, 1988).

In the agriculture belt of northern Kazakhstan and adjacent portions of West Siberia, dust storms also frequently damage crops. In this region, often referred to as the "Virgin Lands," large areas were planted to wheat in the 1950s that are productive in climatically favorable years, but which in years with dry springs can be subject to severe dust storms. Particularly serious ones occurred in 1963, 1965, 1968, and 1974, with the Pavlodar, Omsk, Novosibirsk, and Altay Krai regions being the hardest hit (Sazhin, 1988).

Responsibility for preventing and abating soil erosion lies with the USSR Ministry of Agriculture. As in other environmental problem areas, special directives have come forth from the Council of Ministers, such as the 1975 resolution "On Measures to Improve Organizational Work for Protecting Soil from Wind and Water Erosion." But experience has shown that such directives may not necessarily be effective, particularly if their success depends on significant levels of funding or the diversion of resources from other ministry priorities.

Over the past two decades, though, some success may have been realized, for with the introduction of better soil conservation measures into the Virgin Lands areas in the 1970s, the incidence of serious dust storms has somewhat decreased. However, perhaps in part because of past deflation, crop yields in this region have also generally decreased since the late 1950s, although cyclical climatic patterns may play a role of unknown magnitude here.

Among the techniques used in the efforts to combat deflation are instituting more soil-protective methods of cultivation, and planting rows of moisture-conserving trees (termed *shelterbelts*) around the periphery of the fields (see Figure 7.2). Unfortunately, shelterbelts tend to perform best in the more humid areas where dust storms are inherently less frequent. For reasons that are not clear, shelterbelt plantings have decreased from 44,000 ha in 1980 to 39,000 in 1985, and to only 29,000 in 1987 (*Narodnoye*, 1988, p. 562). Preliminary goals for the 1990s include anti-erosion efforts on perhaps as much as 110 million ha, but no specifics of such a plan are as yet available (*Sovet. Rossiya*, Mar. 26, 1989).

Deflation has also resulted from poorly planned water management projects. The current crisis surrounding the Aral Sea will be examined in chapter 12, but a similar case occurred in the early 1980s that involved the Kara-Bogaz-Gol. The Gol (Gulf) is a large bay on the east side of the Caspian Sea, fed by water flowing into it through a relatively narrow channel connecting it to the Sea. The idea was advanced in the 1970s to dam off the inflow from the Caspian, allow most of the Gol's water to evaporate, and then mine the minerals that had been deposited over time on the floor of the Gol. This would also reduce water losses from the Caspian Sea itself. This was done in 1980, and by 1983 the Gol was completely dry (Nurberdiyev, 1985). But once exposed and dried out, these saline deposits quickly became airborn, falling out (in part) over irrigated agricultural land in the Turkmen republic. Four years after the project began, it became necessary to reverse the decision, and the Kara-Bogaz-Gol was again flooded (Shabad, 1984).

The recent "salt storms" produced by deflation around the Aral Sea have been reported upon by Micklin (1988, 1989) and others; Micklin will discuss the causes and implications of this desiccation in chapter 12.

Reclamation and land use in wetland areas

Discussing any country's "reclamation efforts" is hampered by the fact that the word "reclamation" in English has at least three commonly used meanings. These include (1) the draining of wetlands so that they may be used for some economic activity; (2) the restoration to biological productivity of lands formerly disturbed by mining; and (3) in the United States, the activities of the Bureau of Reclamation which have historically centered around building dams that divert water for irrigation purposes.

In the Soviet Union, the same word (*reklamatsiya*) is also used in a general sense, but more specific words also exist. In this section we will briefly look at wetland drainage (*osusheniye, melioratsiya*) and restoring mined-out lands (*rekul'tivatsiya*); the environmental implications of large-scale irrigation (*orosheniye*) will be reviewed in chapter 12.

The environmental problems associated with economic development in naturally marshy areas are often unappreciated until significant adverse changes have occurred. This tendency can be seen in two Soviet case studies: the West Siberian lowland and the Pripyat Marshes in Byelorussia.

Most of West Siberia is a huge, flat basin drained by the tributaries of the Ob River. Frozen over in winter, the release of large amounts of meltwater during the spring thaw causes the land around the northward-flowing rivers (which may still be blocked by ice near their mouths) to become waterlogged or flooded. In the north, these lands are also underlain by permafrost. Such lands are inherently difficult to develop for economic ends, and little was done with interior West Siberia until the 1970s. At that time, development of the Soviet Union's largest oil and gas deposits got underway on a massive scale in this region, with from 50,000 to 100,000 wells now having been drilled (Volfson and Rosten, 1983).

Two sets of environmental problems emerged. One involves pollution questions; marshy lands in particular are very difficult to clean up once they are contaminated by oil. In 1979, a pipeline rupture under the Ob River was not contained for several days, and in the interim massive amounts of oil poured into the river. There are in addition many other toxic substances used in hydrocarbon extraction, some of which (as well as just plain litter) inevitably find their way into the aquatic environment. The pollution problem is compounded by the fact that, in northern latitudes, rates of biological activity decrease, making natural restitution of ecosystems a much lengthier process.

The other set of problems concerns physical damage to the land. Drilling and the movement of heavy equipment over wetlands can permanently degrade them, with vehicle tracks turning into watercourses in surprisingly little time. In more northerly areas, the underlying layer of permafrost creates additional construction and biosphere management problems. Forests that overlay permafrost can be turned into irreversible marshes once the trees are cut away. Many of these same problems have been encountered in the oil fields of Alaska. In both the US and USSR, realization of the fragility of these northern environments has slowly emerged, but only after considerable damage had been inflicted on the regional ecosystems.

The second example in the USSR involves the extensive lowland areas around the Pripyat River in the southern Byelorussian (White Russian) republic. The Pripyat marshes are best known for the grief they provided the retreating armies of Napoleon. In the postwar Soviet period, it was decided that this vast area could be made agriculturally productive if drainage channels were artificially created to lower the water table in the areas most prone to waterlogging. Much money and effort has gone into this effort, but the results have been discouraging, due to the low productivity of some of the reclaimed areas, the necessity of applying large amounts of chemical fertilizers, and the choice of inappropriate uses for some of the reclaimed lands. In addition, areas of cut or dried-out marshland vegetation in the Pripyat region have at times been subject to wildfires.

Significant degradation to the landscape in this region has also been brought about by the potassium mining operations near the town of Soligorsk. Specific problems which have occurred include salinization of the soil, subsidence and subsequent waterlogging, severe air pollution, mine collapse, loss of agricultural capability, and pollution of local water bodies, all of which has resulted in the need to relocate several whole villages (Petrashkevich, 1988). It may be necessary in this region by the year 2000 to withdraw hundreds of square kilometers of land from agricultural use.

In a study of the optimal uses of the reclamation zone along the Pripyat River, it was noted that a large array of ameliorative measures have "failed to enhance the economic and environmental protective effectiveness of the reclamation projects. This low cost-effectiveness of reclamation work in the [Pripyat marshes] has been a subject for debate at various governmental levels" (Kiselev, 1984, p. 573). This author also notes that a major cause of such problems is that reclamation projects are often begun without preliminary study of the impact on the ecological interrelationships in the region.

Other examples of wetland degradation could be cited. In chapter 12, very serious problems regarding the critically important deltas of the Amu-Darya and Syr-Darya rivers in Central Asia are discussed. The desiccation of these deltas, as well as much of the river courses that lie above them have created enormous economic and environmental problems in this region that will be expensive to correct.

A recent international approach to the worldwide problem of wetland destruction is the Ramsar Convention. This agreement, to which the Soviet Union became a party in October of 1976, commits the participating nations to designate within their boundaries "wetlands of international significance" for the protection of all forms of water-dependent life and ecosystems, and to take appropriate steps to safeguard them. As of 1989, the USSR had designated about twenty areas (mostly lakes and bays, but also the Volga delta) as specially protected wetlands under this convention (Table 11.3).

Table 11.3   *Soviet wetlands protected under the Ramsar Convention*

Wetland	Republic	Area (ha)
Kandalaksha Bay	RSFSR	208,000
Matsalu Bay	Estonia	48,634
Volga Delta	RSRSR	650,000
Kirov Bay	Azerbaidzhan	132,500
Krasnovodsk and North Cheleken Bays	Turkmen	188,700
Sivash Bay	Ukraine	45,700
Karkinitski Bay	Ukraine	37,300
Danube estuary and Yagorlits and Tendrov Bays	Ukraine	113,200
Kurgaldzhin and Tengiz	Kazakhstan	260,500
Lakes of the lower Turgai and Irgiz	Kazakhstan	348,000
Lake Issyk-Kul	Kirgiz	629,000
Lake Khanka	RSFSR	310,000

Source: List of wetlands of international importance, IUCN 1989 edition.

Reclamation of surface-mined areas

The increasing global demand for mineral resources has left large areas of the earth's surface denuded and barren as a result of their extraction. The Soviet Union, as the leading mining country in the world, has more than its share of such scars on the landscape. In 1987, about 43 percent of Soviet coal and 86 percent of its iron ore was being extracted by surface methods (Bond and Piepenburg, 1990). By one estimate, originally appearing in the Soviet journal *Planovoye khozyaystvo*, over 4 million ha of agricultural land have been despoiled by mining wastes (Wolfson, 1988b). By the year 2000, one Soviet specialist estimates that from 950 million to one trillion cu. meters of overburden (overhanging earth or rock) will need to be reclaimed worldwide (Drizhenko, 1985).

The reclamation (meaning restoration to biological productivity) of these mined-out lands is required in most advanced countries (that is, those with the financial resources to accomplish this goal). In the Soviet Union, the 1968 Basic Principles of Land Legislation contained fairly generalized wording that promoted proper care and restoration of disturbed lands (Pryde, 1972, pp. 188 and 199). A more comprehensive statute that specifically stipulates the reclamation of surface-mined lands was enacted in June of 1976. It requires that the fertile

Figure 11.2   Poorly stabilized overburden piles at coal mines near Shakhty

topsoil be conserved for later use when land is first opened up for surface mining, and that the land be restored to support either agriculture or its normal vegetation cover after the mining operations are completed.

Is the law complied with? Apparently, not very well; examples are numerous where "moonscapes" seem to persist in surface mining regions, such as at Soligorsk and in the Donets Basin. In Estonia, where there are several mines extracting oil shale and phosphorite, operations have been sharply criticized by local officials. At the Exhibit of Estonian Economic Achievements in Tallin, the author saw a display illustrating how a waste pile at these mines could be revege-tated and stabilized even while still in use. As it was a drawing, rather than a photograph, it is not clear if this technique is actually in use.

Perhaps the most frequently cited region having problems resulting from hard-mineral extraction is the Donets Basin (Donbass). Ironi-cally, mining in these extensive coal deposits in the eastern Ukraine and nearby areas of the Russian Republic is done mainly by under-ground, not surface, mining. Here, piles of overburden (the layers of material surrounding the mineral deposit that must be removed before the high-value resource can be extracted) have created literal moun-tains on the surface that in places are up to 100 m high (Figure 11.2).

Figure 11.3   The open-pit iron ore mine at the Kursk Magnetic Anomaly

Much effort has been made to vegetate these waste piles, but the task is difficult because this material is generally very infertile (Koutaissoff, 1987, p. 25). As a result, dust from these "mountains" is frequently blown around the countryside, creating problems on the surrounding rich agricultural lands.

Another portion of the country greatly disturbed by mining activities is the Kursk and Belgorod regions, site of the huge "Kursk Magnetic Anomaly" open-pit iron ore mines. This region is the subject of an extensive comprehensive environmental monitoring program (chapter 13), a portion of which involves the adverse effects of the mining operations on the surrounding lands. Although when new mines are opened the topsoil is carefully removed, as required by law, over the numerous years it must wait to be reused it can undergo considerable deterioration, slowly becoming alkaline and turning into a "black powder" (Klyuyev, 1988; Wolfson, 1988c).

The present author had the opportunity to visit one of the giant open pit mines near Kursk in 1983, as part of a delegation studying Soviet environmental problems. The huge excavation was awesome, several kilometers long and over a hundred meters deep (Figure 11.3). We had been assured that a tour of the reclamation operations associated with the open pit mining would be a central feature of our visit. By the end of the day, however, it was clear that such a tour was not forthcoming, and upon inquiring why, it was finally acknowledged that little (if any) land had as yet actually been reclaimed. The reason given was that no portions of the big pit had as yet been inactivated, and thus no land was ready for restoration. This may indeed have been the case. However, it would seem that demonstration sites or test areas would have been created to prove out the ultimate feasibility of the reclamation concept at this location. If such a test site existed, we were not afforded the opportunity to see it.

Hopefully, the 1976 reclamation law is being enforced. But as with so many other environmental problems in the USSR, there may be insufficient resources available to comply fully with the law even if local mining and governmental officials desire to do so, especially under tight *perestroika* budgets. In addition, as poorer mines are brought into development (as is increasingly occurring) the ratio of overburden to useful ore extracted goes up, thereby increasing the volume of spoil banks needing eventual reclamation. As a result, most observers feel that the restoration of land disturbed for mining in the Soviet Union is not highly effective. One recent study concluded that the 1976 Soviet law appeared to be targeted more at restoring lost economic productivity than at implementing a nature protection program (Bond and Piepenburg, 1990).

Managing land resources by the use of *zakazniki*

Chapter 8 reviewed the Soviet system of state nature reserves, or *zapovedniki*, a system which has become fairly well known outside the USSR. But there also exists a companion system of preserved lands called *zakazniki*, about which very little has been published in the West. Yet the *zakazniki* network is actually far larger, taking in over 48 million ha (as compared to less than 20 million for *zapovedniki*). The *zakazniki* system has recently been described in some detail in English translation (Shalybkov and Storchevoy, 1988). Because of its size and importance as a Soviet land management technique, it will be useful to summarize the main features of the system here.

Zakazniki evolved from the old "hunting reserve" tradition of the prerevolutionary nobility. In the Soviet period, they have been defined as a type of semi-preserved territory in which either permanent or temporary limitations are placed upon on-site economic activities, in the interests of natural resource conservation and wildlife reproduction. These restrictions on economic activities are often in place only during certain seasons (such as the breeding season) or for other fixed periods of time (for example, for a few years to allow a particular resource to recover its vitality or productivity). Several categories of *zakazniki* exist, including zoological, botanical, landscape, hydrological, geological, and others.

Zakazniki can be created at either the Union republic or local level. In either case, they are managed by various state agencies which are responsible for natural resource conservation, operating under guidelines put forth in a Model Statute enacted in 1981. In the Russian republic (RSFSR), where they are most widespread, the majority are administered by the agency that controls hunting (Glavokhoty), and regulated commercial hunting is a main feature on many of them. Others, though, may exist specifically to prohibit hunting for a period of time until game reserves become more plentiful. Any land manager (for example, a collective farm) on whose property a *zakaznik* partly or wholly lies is obligated to manage the land within the preserve according to *zakaznik* rules. Unlike *zapovedniki*, *zakazniki* generally have no permanent on-site research facilities or staff. Some that are actively involved in game procurement may have a cadre of professional hunters.

Zakazniki may be multi-purpose (that is, created to protect a number of features) or they may be set up for a single purpose. The majority tend to be of the zoological type, and have been set up to manage hunting, especially commercial hunting, while at the same time to conserve wildlife generally and in particular to expand the numbers of important game species (for example, sable and beaver). Shalybkov and Storchevoy indicate that the number of *zakazniki* actively engaged in propagating rare or endangered species, or in protecting migratory species, is not great. One that does play an important role in this regard, however, is the large preserve established around Lake Khanka (a Ramsar Convention wetland) in the Soviet Far East near Vladivostok. Figure 10.3 indicates the importance of this region for endangered species.

Table 11.4 indicates that as of 1985 there were 2,912 *zakazniki* in the Soviet Union, having a total area of approximately 480,755 sq. km, or a little over 2 percent of the country's total area. Because they are generally located in biologically or geologically significant areas, though, their importance is greater than the 2 percent figure might at first suggest. For example, there are over thirty *zakazniki* in the basin of Lake Baikal (Vorob'yev and Martynov, 1989). Table 11.4 indicates that 41 percent of all *zakazniki*, and 92 percent of their land area, are located

Table 11.4 *Area in zakazniki in the USSR by Union republic*

Republic	Total No.	Total 1,000 ha	Percent of SSR	Zoological No.	Zoological 1,000 ha	Botanical No.	Botanical 1,000 ha	Landscape No.	Landscape 1,000 ha	Other No.	Other 1,000 ha
RSFSR	1,203	43,986.6	n/g	1,049	43,532.4	81	231.4	25	113.7	48	109.1
Ukraine	1,103	424.9	7.0	118	111.5	473	126.0	55	53.5	457	133.9
Belorussia	50	606.7	2.9	6	398.3	33	169.9	4	1.5	7	37.0
Uzbek	5	78.6	0.2	5	78.6						
Kazakh	37	408.1	0.2	15	223.6	22	184.5				
Azerbaijan	16	262.8	3.0	16	262.8						
Lithuania	174	185.0	2.8	25	3.9	56	10.4	63	103.7	30	67.0
Georgia	1	10.0	0.1							1	10.0
Moldavia	11	3.2	0.1	2	0.4	9	2.8				
Latvia	148	104.0	0.2	6	12.9	29	1.4	37	36.2	76	53.5
Kirgiz	66	309.2	1.6	16	289.9	31	11.7			19	7.6+
Tadzhik	13	800.6	5.6	11	776.7	2	23.9				
Armenia	16	85.1	2.9	7	51.6	4	6.3			5	27.2
Turkmen	12	607.0	1.2	5	455.0	7	152.0				
Estonia	57	203.7	4.5	2	1.6	9	4.7	14	80.2	32	117.3
USSR Total	2,912	48,075.5		1,283	46,199.1	756	925.0	198	388.8	675	562.6

Source: Shalybkov and Storchevoy, 1988 p. 598.

in the RSFSR. Of those in the RSFSR, 34 percent are administered at the republic level, and the rest at a more local level.

The ability of the *zakazniki* system to achieve its conservation goals has been constrained:

The limited effectiveness of *zakazniki*, both for the goal of conserving nature as well as for [promoting] the hunting economy, is caused by the absence of land use laws for the various types of preserved territories . . . (T)he 1980 Law of the USSR on the Protection and Utilization of the Animal Kingdom, states in Article 25 that "On *zakazniki* and on other specially protected territories, the implementation of particular types of uses of the animal kingdom, and other activities, may be restricted or completely prohibited, if they are incompatible with the goal of protection of the animal kingdom." However, legislation has thus far not set up workable measures for securing these restrictions. A prohibition on hunting cannot secure the conservation of large numbers of animals, if on a given territory undesirable changes are occurring in the places they inhabit. (Shalybkov and Storchevoy, 1988, pp. 592–3)

Although zoological *zakazniki* are the most widespread overall, in the majority of Union republics outside the RSFSR, botanical, landscape, and other *zakazniki* outnumber zoological ones. In the Ukraine, Lithuania, Moldavia, Latvia, and Estonia, even their total area is greater. However, over the country as a whole, the average size of the hunting *zakazniki* is far greater than those in the other categories; the hunting *zakazniki* average 36,000 ha in size, the others only 1,150.

It is clear that the *zakazniki* play somewhat different roles in the various republics. The Georgian and Lithuanian republics are about the same size, but Lithuania has 174 of these preserves and Georgia has 1 (in fairness, however, Georgia has far more *zapovedniki* than does Lithuania). The Russian republic and some others use them primarily for game management purposes, whereas the Baltic republics use them mainly to preserve specific features of the landscape. The suggestion that they do not always fulfill their conservation mandates as well as they should can probably be largely traced to inadequate funding and personnel. Even so, their importance in the Soviet conservation effort is clear, and in the face of growing population and economic pressures, they have the potential to become even more valuable in the future.

In the past, the vast land resources of the Soviet Union were viewed basically as a cornucopia from which the tempting fruits could be extracted forever in whatever manner was most efficient. Soil erosion and a general reduction in biotic productivity was the result. Today, there is abundant evidence of the errors in the old way of thinking, and many works dealing with proper management of the land have been published (e.g., Nosov *et al.*, 1986). Soviet scientists generally know how to properly manage the resources of the land, but instilling this wisdom throughout the planning and ministerial establishment, and obtaining the necessary budgetary support, is more difficult. It is to be hoped that the latter goals can become an integral part of the *perestroika* reforms.

12 The water crisis in Soviet Central Asia*

An abundant and assured supply of fresh water is an economic and social necessity to modern societies. Water withdrawals are particularly large in arid regions where irrigation has been extensively developed. However, since irrigation is a heavily consumptive use of water (i.e., a large proportion of water withdrawn is not returned), sources of supply such as rivers and ground water suffer significant depletion with attendant ecological, economic, and social consequences. The management of consumptive water development in arid environments, which includes meeting both economic and ecological demands, is a complicated and difficult task.

The most serious water management problem in the USSR occurs in Soviet Central Asia, a region situated in the arid zone of the country. Intensive development of irrigation here has depleted river flow and led to the drying of the Aral Sea, a huge saline lake, with accompanying severe impacts. The Soviet government until recently proposed to respond to the region's water management problems by large-scale, long-distance water importation from Siberian rivers far to the north. This project has been halted and local means of resolving water supply problems are being pursued. These, however, may fail to adequately resolve the water crisis.

Water availability and usage in Central Asia

Central Asia lies among the deserts of the extreme south-central part of the Soviet Union, between the Caspian Sea on the west and the high chains of mountains on the south and east. Central Asia is defined in this chapter to include not only the Uzbek, Tadzhik, Kirgiz, and Turkmen Soviet socialist republics (SSRs) but the two southern oblasts

* This chapter was prepared by Philip P. Micklin, Department of Geography, Western Michigan University, Kalamazoo, Michigan. Research support provided by the National Council for Soviet and East European Research, the National Academy of Sciences, and the Lucia Harrison Fund, Department of Geography, Western Michigan University.

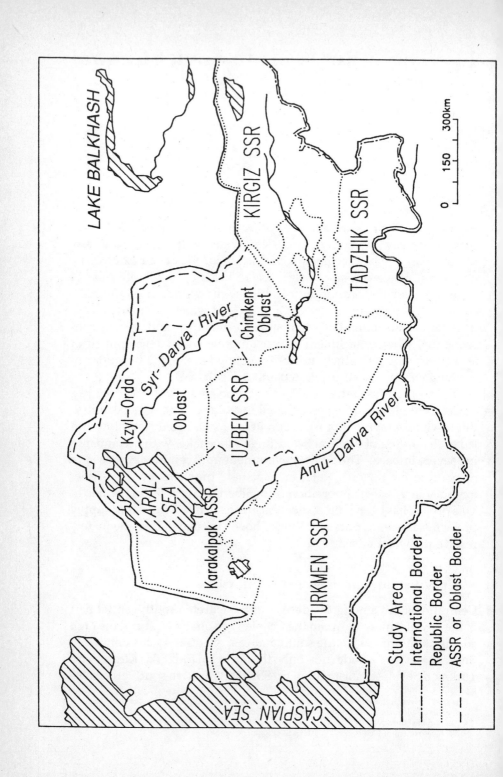

LAKE BALKHASH

KIRGIZ SSR

TADZHIK SSR

0 150 300km

Syr- Darya River

Chimkent Oblast

Kzyl–Orda Oblast

UZBEK SSR

Amu- Darya River

ARAL SEA

Karakalpak ASSR

TURKMEN SSR

CASPIAN SEA

—— Study Area
–·– International Border
····· Republic Border
········ ASSR or Oblast Border

(provinces) of the Kazakh SSR (Kzyl-Orda and Chimkent) which are also found in the Aral Sea drainage basin and have a commonalty of water management problems with the four republics (Figure 12.1). Central Asia's population was 35 million on January 12, 1989, 12 percent of the USSR total. The rate of natural increase (births minus deaths) averaged 2.54 percent over the inter-censul period 1979–89 compared to a national rate of 0.87 percent. This is the most rapid growth of any region in the USSR and exceeds the rates in many developing countries. High fertility, the young age structure, and minimal out-migration, make rapid population growth for Central Asia a near certainty well into the next century. Thermal conditions for plant growth are the best in the USSR: sufficient to raise heat-loving crops such as grain corn, sorghum, rice, and soy over much of the region and sufficient for cotton cultivation in the desert and plains and foothills in all but its northern part.

The surrounding mountains capture plentiful moisture and store it in snow fields and glaciers whose run-off, primarily during spring thaw, feeds the region's rivers. Estimated average annual river flow in Central Asia is 122 cu. km/yr (Voropayev *et al.*, 1987, 182f, 226ff). The Aral Sea drainage basin accounts for 90 percent (110 cu. km/yr) of this. It includes the two largest rivers, Amu-Darya (73 cu. km/yr) and Syr-Darya (37 cu. km/yr), in Central Asia. Discharge is maximum where the rivers exit the mountains but decreases rapidly as they cross the deserts. The Amu and Syr-Darya (until the 1960s) lost around half their flow before reaching the Aral Sea. Usable supplies of ground water are estimated at 18 cu. km/yr. Thus, aggregate average annual water resources for Central Asia are around 140 cu. km/yr, with 90 percent found in the Aral Sea drainage basin.

Water use in Central Asia is great. For 1980, withdrawals were estimated at 134 cu. km with consumptive use (water directly lost to evaporation and transpiration or incorporated into plants, animals, or other products) of 80 cu. km (60 percent of withdrawals) (Voropayev *et al.*, 1987, pp. 212–15). The balance of 54 cu. km (40 percent of withdrawals) constituted return flows from sources such as leakage from canals and pipes, surface and ground water run-off from water applications, water collected by irrigation drainage systems, and end discharges of canals or pipes. A significant portion of return flows, primarily from water used for irrigation, is subsequently evaporated or transpired (i.e., consumptively used) rather than returning to rivers or adding to the ground water reservoir. Consumptive use and reservoir evaporation (together totalling 91 cu. km) were 74 percent of

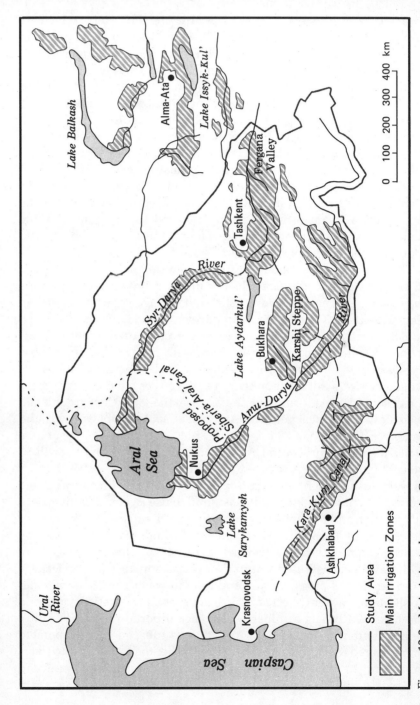

Figure 12.2 Main irrigated areas in Central Asia *Source:* Adapted from original map by P. P. Micklin

average annual flow and 65 percent of average annual water resources in Central Asia. For the Aral Sea basin, the figures were 81 percent and 71 percent. However, since a significant portion of return flows (probably around 40 percent) are lost to evaporation and transpiration, total losses from river flow are considerably greater. Indeed, anthropogenic factors had reduced the flow of the Amu and Syr-Darya into the Aral Sea to near zero by the 1980s (Micklin, 1988).

Irrigation is the dominant user of water in Central Asia, accounting in 1980 for 90 percent of withdrawals, 95 percent of consumptive use (excluding reservoir evaporation), and 84 percent of return flows (Voropayev et al., 1987, pp. 212–15). The fundamental cause of the water crisis in Central Asia is irrigation; water use by other sectors is, by comparison, insignificant. Even small percentage water savings in irrigation could provide enough water to meet future needs of other economic and social sectors in Central Asia.

Central Asia is the most important region of irrigation in the USSR (Figure 12.2). Irrigation has been practiced there for thousands of years. By 1984, state-run irrigation systems encompassed 7.2 million ha, 38 percent of the national total. The USSR is the world's third largest cotton producer. All Soviet cotton is irrigated and 95 percent is raised in Central Asia. This is the region's pre-eminent crop. Irrigation is also crucial to food and fodder production: 40 percent of the USSR's rice, one-third of its fruit and grapes, and a quarter of its vegetables and melons are grown on irrigated lands here. Irrigation provides over 90 percent of crop production in Central Asia (Vasil'yev, 1987).

Given the near certainty of continued rapid population growth in Central Asia for the foreseeable future, rapid expansion of the regional economy and of food production is essential. Even now per capita consumption in a number of basic foodstuffs is significantly below national averages (Dukhovnyy and Razakov, 1988). Irrigation as a means to meet Central Asia's growing employment and food needs has much to recommend. It is labor intensive and does not require a highly skilled work force, which is also true of related industries such as food processing, cotton textiles, and clothing manufacture. Crops can be grown without irrigation on the more humid slopes of the region's foothills and mountains and hardy livestock (sheep, goats, and camels) can even be pastured in the desert. But crop yields and meat production are much lower and more variable than with irrigation. Until recently, continued growth of irrigation in Central Asia was assumed. The long-range reclamation program approved in 1984 projected the irrigated area in Uzbekistan, Turkmenia, Kirgizia, and

Tadzhikistan to rise 23–27 percent between 1985 and 2000 (*Pravda*, Oct. 27, 1984, pp. 1–2). Because of greater emphasis by the Gorbachev regime on alternative means of increasing agricultural production and, particularly, the dire water supply situation in Central Asia, planned irrigation expansion here has been scaled back (Vasil'yev, 1987).

Many Soviet water management experts believe the water resources of the Amu and Syr-Darya drainage basins reached full utilization in the early 1980s owing to heavy withdrawals for irrigation. An authoritative 1987 study of water management in the USSR concluded that total withdrawals for irrigation were around 100 cu. km in these two river basins with 35 cu. km lost to filtration ("Concerning . . . "). Of total filtration losses, 14 cu. km was ultimately consumptively used, suggesting 21 cu. km (35 − 14), or 21 percent of withdrawals, constituted water returned to rivers and 79 cu. km or 79 percent overall consumptive use. The Syr and Amu-Darya are considered to have the most strained water balances of the USSR's major rivers.

Water management and irrigation in Central Asia have become controversial. The most strident critics contend that water wastage in irrigation here is enormous, and maintain that large amounts of water can be freed without great difficulty or cost that will suffice to meet regional needs far into the future (Micklin and Bond, 1988). Water management specialists, on the other hand, are more cautious, predicting that the amount of water that can be freed is much less than the optimists believe, that it could somewhat alleviate but by no means "solve" regional water problems, and that implementing the necessary water saving measures will be lengthy, costly, and complicated.

There is general agreement that, first and foremost, water use efficiency in irrigation must be improved. A rough estimate for the Aral Sea basin, which has some 97 percent of the irrigated area in Central Asia, is that by the early 1980s only 60 percent of withdrawals arrived at the field with 40 percent (perhaps equalling 50 cu. km) lost in the conveyance system (*Pravda*, Apr. 14, 1988, p. 3). A more accurate measure of water use efficiency in irrigation is the portion used productively by crops. This figure may have been as low as 44 percent in the late 1970s in the basins of the Amu and Syr Darya. However, not all of the residual 56 percent was lost for further use: one quarter to one third constituted return flows which re-entered the river network by surface or ground water routes. Nevertheless, it was clear by the early 1980s that there were major opportunities for improving the efficiency of irrigation in Central Asia and thereby freeing sizable amounts of water for expanding irrigation and for other purposes. Soviet water management experts have set the average efficiency target for irrigation systems in Central Asia around 80 percent. This refers to the percentage of withdrawn water arriving at the fields. Assuming 1980 withdrawals for irrigation were 120 cu. km, this level of improvement would have allowed irrigating the same area with 30 cu. km lower withdrawals (Voropayev et al., 1987, pp. 212–15).

The improvement of irrigation efficiency in Central Asia has been a priority since 1982 when water scarcities caused strict limits to be placed on water consumption. It has received additional emphasis under Gorbachev as part of the general program for agricultural intensification and has been given the highest priority by the Ministry of Reclamation and Water Management for the 12th five-year plan (Vasil'yev, 1987). The effort to raise irrigation efficiency has had results in Central Asia. One source states that average per hectare withdrawals fell from 18,700 cu. m/ha in 1980 to 13,700 cu. m/ha in 1986 (Dukhovnyy and Razakov, 1988). This allowed around a 20 percent increase in the irrigated area with a decrease in water withdrawals.

There are a variety of measures and strategies to improve the water use situation in Central Asia. Rebuilding of older irrigation systems to reduce water losses has received the greatest stress. Given the long history of irrigation development here, a larger share of the irrigated area is served by antiquated and inefficient irrigation systems than in any other part of the USSR. Thus, 70 percent of new irrigation investment here during the 12th five-year plan was devoted to the renovation effort (Dukhovnyy and Razakov, 1988).

Reconstruction involves implementation of a complex of measures. Of fundamental importance is reducing losses from canals by lining them with concrete or other coverings, and by shortening their length through consolidation. Installing or improving collector-drainage facilities to remove excess ground water and prevent water logging and soil salinization is also crucial. Most of the older irrigation systems in Central Asia either lack engineered drainage networks entirely or have crude open channels, frequently choked with weeds, that do not effectively remove water. Proper drainage not only prevents water logging and soil salinization but also lowers water usage by greatly reducing the amount of water needed to flush salts from the soil.

Improving water application at the field is another essential measure for raising the efficiency of water use. In the early 1980s, 98 percent of irrigation in Central Asia was by surface methods, where water flows from a canal or flume directly onto the fields, mostly through furrows. The efficiency of furrow irrigation is generally low but can be raised substantially. There are also more modern and efficient irrigation technologies such as sprinklers, drip, and intersoil whose use could be expanded to some degree in Central Asia, though each has limitations. Consequently, surface irrigation will remain the most important water application technology here.

Automation, computerization, and telemechanization of large irrigation systems are being implemented to improve water usage. The basis of the program is the installation of water measurement and regulating devices along main and distributary canals that can transmit data to and be directed by a central facility. Centralized water management authorities were established for the Amu-Darya and Syr-Darya basins in 1988. The systems are supposed to gather and process information on estimated water supply and needs within the basins, determine an annual water allocation plan, control the distribution of water, and keep track of actual water allocations, uses, and conditions.

Another water conservation measure being implemented is a shift from high water consuming crops such as cotton and rice to lower users such as vegetables. This change will not only lower per hectare withdrawals but contribute to the improvement of regional food supplies. Nevertheless, cotton will continue to be the dominant irrigated crop in Central Asia for the foreseeable future because of its great economic importance to the region. Efforts are being made to raise the yield per cubic meter of applied water of all crops and to refine irrigation water application standards to adjust them more precisely to crop water consumption requirements.

There are opportunities to develop new or currently under-utilized water resources for irrigation in Central Asia. Ground water usage could perhaps be expanded to 17 cu. km/yr without adverse effects on surface water flow (Voropayev *et al.*, 1987, pp. 182–83). A larger share of irrigation drainage water might be reused but much of it is so saline that without dilution it would damage most crops (Chernenko, 1986). The system of reservoirs in the basins of the Amu and Syr-Darya when finished will allow the increase of available water resources from these rivers, compared to natural conditions, by 15 cu. km during low flow years but will lose large amounts of water to evaporation and have adverse ecological effects. Finally, it is possible to utilize periodic run-off collecting in ephemeral streams and clay-pan basins for small-scale irrigation.

On the economic and institutional side, water pricing is the most promising means of improving water use in irrigation. Water used for agricultural purposes is provided free which promotes inefficiency and waste. Although charges for irrigation water have been employed at times in the Soviet Union, the system now under discussion goes far beyond past measures. It would force state and collective farms to pay for water delivered to them by republic water management agencies, who would be charged for their water withdrawals from rivers and ground water. Although introduction of a price for irrigation water in principle would be beneficial, implementation is beset by problems such as a lack of water-delivery measurement facilities, and, most importantly, the out-dated and illogical system of agricultural prices. Consequently, irrigation water pricing was being introduced on an experimental basis in only three oblasts, including Tashkent in Uzbekistan, during 1988–89. Based on this experience, it is planned to begin implementing irrigation water charges nationally in 1991 (*Pravda*, Sept. 26, 1988, p. 4).

The Aral Sea problem

If the water management difficulties of Central Asia related to irrigation were not enough, there is the added complication of the desiccation of the Aral Sea. Situated in the heart of Central Asia, this huge, shallow, saline lake has no outflow; its level is determined by the balance between river and ground water inflow and precipitation on its surface on the one hand, and evaporation from the sea on the other. The sea over the past 10,000 years has experienced major recessions, initially caused by natural factors but also resulting from human activities during the last several millennia (Kes', 1978). For the period from 1910, when accurate and regular level observations began, to 1960, the Aral underwent level changes of less than one meter. However, during the past twenty-nine years the sea's surface has dropped precipitously (Figure 12.3). In 1960, the sea's level was at 53.4 m, its area was 68,000 sq. km, volume 1,090 cu. km, average depth 16 m, and average salinity near 10 grams/liter (Micklin, 1988). The Aral was the world's fourth largest lake in area. By 1987, sea level had fallen 13 m, the area had decreased by 40 percent, volume diminished by 66 percent, average depth declined to 9 m, and average salinity risen to 27 g/l. The sea had dropped to sixth place in area among the world's lakes.

The recent recession and its associated impacts have been the most rapid and pronounced in 1,300 years (Kes', 1978). The severe and wide-

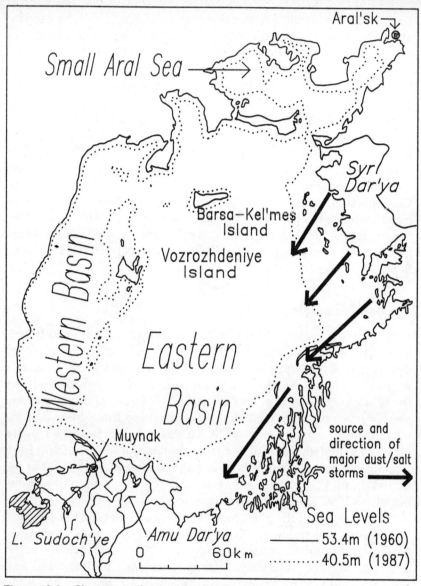

Figure 12.3   Changes in the Aral Sea level *Source:* Map provided by P. P. Micklin

spread ecological, economic, and social consequences are steadily worsening. The Aral situation has been characterized in the Soviet press as "one of the very greatest ecological problems of our century," and compared in magnitude to the Chernobyl nuclear accident and the Armenian earthquake (*Pravda Vostoka*, Sept. 10, 1987, p. 3 and June 2, 1989, p. 1).

As in the past, the cause of the modern recession of the Aral is marked diminution of inflow from the Syr and Amu-Darya, the sea's sole sources of surface inflow, which has increasingly shifted the water balance toward the negative side. Excepting the heavy flow year of 1969, river discharge to the sea has tended to go steadily downward since 1960. Evaporative losses from a shrinking water body diminish as its area decreases, forcing the water balance toward equilibrium. Hence, in the future, the Aral's level, supported by residual irrigation drainage and ground water inflow, should stabilize. However, this is not likely to occur for many years since the difference between inflow and net evaporation (evaporation minus precipitation) is currently large and negative. If present processes are allowed to continue unchecked, the Aral will be reduced to several briny remnants in the next century.

The causes of reduced inflow since 1960 are both climatic and anthropogenic. A series of dry, naturally low flow years occurred in the 1970s and 1980s, but the main factor reducing river flow has been large consumptive withdrawals, mostly for irrigation. Although irrigation has been practiced in the basins of the Amu and Syr-Darya rivers for millennia, the consumptive use of water by irrigation prior to the 1960s did not measurably reduce inflow to the Aral. This usage was compensated by correspondingly large reductions of natural evaporation, transpiration, and filtration, particularly in the deltas of the Syr and Amu-Darya, owing primarily to truncated spring flooding. Also, the installation of drainage networks increased irrigation return flows to these rivers, albeit of reduced water quality (Micklin, 1988).

Factors that compensated the earlier growth of consumptive withdrawals reached their limits in the 1960s. Hence, as irrigation in the Aral Sea basin expanded from around 5 to over 7 million ha over the past three decades, the resultant consumptive water use roughly doubled from 40 to nearly 80 cu. km, but the increase in water usage has not been balanced by commensurate reductions in natural losses (Micklin, 1989, pp. 57–58). The disproportionate increase of consumptive use (compared to the size of the newly irrigated area) resulted from the development of new areas such as the Golodnaya (Hungry) steppe along the Syr-Darya (Figure 12.2), where huge volumes of water went to saturate dry soils and drainage water flowed into the desert or natural depressions to evaporate rather than returning to rivers. More water was used for increased soil flushing needed to combat secondary salinization. The creation of new reservoirs also contributed to losses of river flow, through the filling of dead storage and increased evaporation from their surfaces.

The Kara-kum Canal has been the single most important factor contribution to the diminution of inflow to the Aral in recent decades. The largest and longest irrigation canal in the USSR, it stretches 1,300 km westward along the southern margins of the Kara-Kum Desert from where the Amu-Darya emerges from the mountains (Figure 12.2). Between 1956 and 1987, 225 cu. km of water were diverted into it from the rivers as annual withdrawals rose from less than 1 cu. km to more than 14 cu. km (Micklin, 1988). All of the water sent along the Karakum Canal is lost to the Aral.

When plans for a major expansion of irrigation in the Aral Sea basin were developed in the 1950s and 1960s, it was anticipated that this would reduce inflow to the sea, substantially reducing its size. At the time, a number of water management and desert development experts believed this a worthwhile trade off: a cubic meter of river water used for irrigation, they calculated, would be more economically beneficial than the same volume delivered to the Aral Sea. Although a small group of scientists warned of serious negative affects from the Aral's desiccation as early as the mid-1960s, they were not heeded (Micklin, 1988; Ellis, 1990). Time has proven the more cautious scientists not only correct but conservative in their predictions. Several of the most important impacts are discussed briefly below.

The Aral contained an estimated 10 billion metric tons of salt in 1960. As the sea has shrunk, exposing some 27,000 sq. km of former bottom, enormous quantities of salts have accumulated. Owing to the salt concentrations, the former bottom is proving stubbornly resistant to natural and artificial revegetation. Consequently, the blowing of salt and dust has become a severe problem. The largest plumes arise from the broad exposed strip along the sea's northeastern and eastern coast and extend for 500 km (Figure 12.3). Major dust-salt storms were first spotted by Soviet cosmonauts in 1975, and by 1981, twenty-nine large storms had been identified by Soviet scientists from analysis of satellite imagery. The majority of storms moved in a southwest direction which carried them over the ecologically and agriculturally

Figure 12.4   Kara-Kum Canal on outskirts of Ashkhabad *Source:* Photograph courtesy of P. P. Micklin

important delta of the Amu-Darya. More recent observations by Soviet cosmonauts indicate the frequency and magnitude of the storms is growing as the Aral shrinks; an especially severe storm occurred April 29, 1989. Around 43 million metric tons of salt are carried annually from the sea's dried bottom into adjacent areas and deposited as aerosols in rainstorms and dew over 150,000–200,000 sq. km. The deflated salts, particularly sodium chloride and sodium sulfate, are toxic to plants (Bel'gibayev, 1986).

As the sea has shallowed, shrunk, and salinized, aquatic productivity has rapidly declined. By the early 1980s, twenty of twenty-four native fish species had disappeared, twelve of which had commercial importance, as the fishery harvest (48,000 metric tons in 1957) fell to zero (Micklin, 1988). By 1990, the number of endemic fish had dropped to zero (Ellis, 1990, p. 85). Major fish canneries at the former ports of Aralsk and Muynak have slashed their work force and barely survive on the processing of high-cost fish from distant oceans. Employment directly and indirectly related to the Aral fishery, reportedly 60,000 in the 1950s, has disappeared. The demise of commercial fishing and other adversities have led to an exodus from Aralsk and Muynak and the abandonment of former fishing villages.

The Aral's shrinkage and the greatly reduced flow of the Syr and Amu-Darya have devastated these rivers' deltas. Thirty years ago, the deltas possessed great ecological value and provided livestock pasturage, spawning grounds for commercial hunting and trapping. These uses have been lost or severely degraded. For example, the area of *Tugay* forests, composed of dense stands of water-loving plants mixed with shrubs and tall grasses fringing delta arms and channels to a depth of several kilometers, had been halved by 1980. Disappearance and degradation of vegetational complexes and water table declines have contributed to desertification in both deltas, and greatly reduced the numbers and diversity of deltaic fauna.

Several other major adverse consequences of the Aral's recession are also apparent. Climate around the Aral has become more continental with warmer summers, cooler winters, lowered humidity, and a shortened growing season, which has forced some cotton plantations to switch to rice. The flow of artesian wells and the level of ground water has dropped all around the sea, leading to dried wells and springs and the degradation of natural plant communities, pastures, and hay fields. The reduction of river flow, the salinization and pollution by agricultural, industrial, and urban effluents of the remaining flow, and the lowering of ground water levels has caused drinking water supply problems. Drinking water contamination is believed to be the main cause of high rates of intestinal illnesses, hepatitis, kidney failure and liver ailments, throat cancer, birth defects and even typhoid and cholera (*Pravda Vostoka*, June 3, 1989, p. 2). The infant mortality rate in areas adjacent to the sea is four times the national figure (Lushin, 1988). Desert animals drinking from the Aral Sea are dying because of its greatly increased mineralization, including the endangered kulan (Asiatic wild ass) and saiga (steppe antelope) that live in the *zapovednik* on Barsa-Kelmes Island.

There are no accurate figures on damages associated with the Aral's recession but a 1979 study concluded that aggregate damages within the Uzbek republic totaled 5.4 to 5.7 billion rubles. Two water management experts have cited 100 million rubles per year as the "social product" losses in the Amu-Darya delta (Dukhovnyy and Razakov, 1988). A popular article listed, without elaborating, 1.5 to 2 billion rubles as the annual losses for the entire Aral Sea region (Micklin, 1988).

Possible approaches to resolving the problem

If measures are not taken to deliver more water to the Aral, it will continue to shrink and in the next century become a set of lifeless, residual brine lakes. Restoring the sea to its pre-1960 dimensions would take an average annual inflow of at least 50 cu. km, ten times the river inflow for 1981–85, and is clearly out of the question. However, a smaller Aral Sea with some economic and ecological importance could be preserved with considerably less inflow. A number of schemes to accomplish this as well as to ameliorate environmental conditions around the sea have been proposed. The simplest and quickest approach would be to supplement the sea's water balance by channeling irrigation drainage water to it that is now lost to evaporation or accumulated in lakes. Perhaps 10–12 cu. km of drainage water annually could be sent to the Aral by collectors paralleling the Amu and Syr-Darya rivers which would also serve to keep this pesticide, herbicide, and defoliant-laden, saline flow out of the two rivers (Micklin, 1988). However, unless treated before release to the Aral, it would further degrade the sea's water quality. Work on a 1,500 km collector for the Amu-Darya is underway but will take years to complete (*Izvestiya*, Sept. 27, 1987, p. 2). Diverting irrigation drainage water to the sea will dry the two largest lakes supported from this source, Aydarkul' and Sarykamysh (Figure 12.2). With an aggregate area over 5,000 sq. km, they have developed considerable fishery and wildlife habitat significance which will be lost (although this will occur inevitably anyway as the lakes continue to accumulate salt and toxics contained in irrigation drainage).

Delivery of 12 cu. km of irrigation drainage water plus 4 cu. km of net ground water inflow would support a sea of only 20,000 sq. km whose salinity would be so high (well above 50 g/l) that its ecological and economic value would be minimal. Thus, additional measures are necessary to save the Aral and to reduce the adverse impacts of its recession. For example, the sea could be partitioned with dikes, and low salinity preserved in those parts which received river inflow (Micklin, 1989, pp. 69–71). Several projects of this genre have been put forward by Soviet experts requiring minimum annual surface inflows from 8 to 30 cu. km and maintaining an "active," low salinity water body of 12,000 to 30,000 sq. km.

Another approach is to focus on improving the ecologically and economically vital deltas of the Amu and Syr-Darya. For the former, the plan is to construct a dike on the dried bottom facing the sea to create a shallow reservoir that would raise water levels in the delta, allowing partial restoration of its former ecological and economic value. The plan would require 12 cu. km of water per year, mainly from irrigation drainage. The dried sea bed in front of the delta would be stabilized by plantings. The residual Aral Sea would stabilize near the 30 m level (10 m lower than in 1987) and have a salinity of 60 g/l. Estimated project cost is 406 million rubles. A similar plan could be implemented for the Syr-Darya delta, requiring some 7 cu. km/yr.

More water could also be provided to the Aral from the rivers of Western Siberia lying to the north. A structural trough (the Turgay Gate) with a maximum elevation of 120 m links the arctic and Aral Sea drainage basins. Bringing Siberian water to the Aral, although ecologically disruptive, expensive, and a major engineering feat, is technically feasible (Micklin, 1989, pp. 71–81). The diversion possibility was recognized even in Tsarist times, but the first serious schemes were formulated during the Soviet era, including the enormous Davydov project that would have flooded a huge portion of western Siberia, caused tremendous ecological and economic harm, and cost several hundred billion rubles in today's currency. (A review of 1960 era schemes appeared in the November 1972 issue of *Soviet Geography*).

The 1970s was a period of intensive development of water redistribution plans but with a greater focus on minimizing their potential environmental impacts. By the end of the decade, detailed designs had been formulated for diversions in both the European and Siberian parts of the country. The first phase of a Siberian transfer was undergoing detailed engineering design in 1985 and was scheduled for implementation by the late 1980s or early 1990s (Micklin, 1986). It would take 27.2 cu. km annually from the Ob' and Irtysh rivers in Western Siberia (see Figure 1.1). Water would be sent 2,500 km southward through the Turgay Gate, into the Aral Sea basin, and as far south as the Amu-Darya by a system of low dams, pumping stations, and a huge earth-lined "Sibaral" (Siberia to Aral) canal (Figure 12.2). Providing more water for irrigation was the scheme's main purpose (90 percent was intended for this sector) but it would have helped stabilize the Aral as well, for example, by increasing irrigation return flows to the Amu and Syr-Darya rivers.

During the Gorbachev era, the fortunes of the European and Siberian schemes have waned. The concept of north-south water transfers has been bitterly attacked in the Soviet popular media as ill-conceived, poorly planned, enormously expensive, and environmentally harmful (Micklin and Bond, 1988; Micklin, 1987). In August 1986, a governmental decree halted construction on the European project and planning for the Siberian undertaking. However, research on the scientific, ecological, and economic problems associated with interbasin water transfers was to continue. A perception of excessive costs compared to expected benefits appears to have been the dominant factor in the cancellation of the projects (Micklin, 1987).

Nevertheless, the Siberian water transfer project may be revived, if, as is likely, regional water resources prove inadequate to meet future economic and social needs and preserve a viable Aral Sea. Central Asian water management experts are again making a case for water transfers from the north as the only means to save the region from a catastrophe (Micklin, 1989, pp. 79–81). The January 19, 1988, decree of the Party Central Committee and Council of Ministers on improving water use in the country directed that scientific study of north-south water transfers continue ("Concerning first priority measures," 1988). Reportedly, Gorbachev during his April 1988 visit to Uzbekistan, after pleas from local officials, agreed to a new feasibility study of the Siberian project (*The Guardian*, Apr., 24, 1988, p. 8).

One possibility could be a scaled-down version of the Siberian plan in which the diverted water would be intended specifically for the Aral and not for irrigation expansion. Water could be delivered to the northern part of the sea, shortening the route and reducing costs. The project could be defended as necessary to save the Aral – arguably a benefit outweighing any environmental harm caused in western Siberia. Ten to fifteen cu. km of fresh water from Siberia per year and implementation of one or a combination of local measures could probably preserve the Aral near its present size (40,000 sq. km), and lower salinity to ecologically tolerable levels.

A national rescue effort has been mounted to save the Aral Sea (Micklin, 1989, pp. 81–90). Under pressure from the popular media, concerned scientists, and a "Save the Aral Committee" organized by the Uzbek Writers' Union, the government issued a decree on September 19, 1988, to implement a program to ameliorate the Aral problem. Based on recommendations from an expert commission, it directed that efforts be implemented between 1988 and 2000 to preserve the sea itself as well as improving the deteriorating ecological, drinking water supply, and human health conditions in adjacent areas. Minimum guaranteed inflow to the Aral is to steadily increase from 8.7 cu. km in 1990 to 21 cu. km in 2005 from water savings realized through efficiency improvements in irrigation and the delivery of irrigation drainage water. The Amu and Syr-Darya deltas are also to be preserved by schemes similar to those discussed above, and a plan developed to stabilize the dried sea bottom.

The decree is being taken seriously. Provisions have been made to guarantee financing, work on some elements of the plan (e.g. delivering more drainage water to the sea and improving drinking water supplies and medical services for the population) is underway, and oversight agencies to ensure compliance of responsible organizations have been established (*Pravda Vostoka*, May 14, 1989, p. 2). Nevertheless, vehement complaints are being heard from Central Asians that the program is going slowly and poorly. Tulepbergen Kaupbergenov, the well-known Karakalpak writer, scathingly criticized current efforts in his speech to the Congress of Peoples' Deputies on June 2, 1989 and stated that during the first five months of 1989 not one drop of river water had reached the Aral (*Pravda Vostoka*, June 3, 1989, p. 2).

Furthermore, even if the program is fully implemented in a timely fashion, results may be disappointing. The stepped increase of minimum guaranteed annual inflow to the Aral to 21 cu. km/yr by 2005 will result in the sea's continued shrinkage, albeit at a decreasing rate, into the next century. The sea would stabilize 5 or 6 m below the current level of 40 m and have an area around 28,000 sq. km compared to 40,000 sq. km today. Unless measures are taken to desalinate and detoxify irrigation drainage water delivered to the sea, average salinity will not only be above the ocean's but the Aral will be seriously polluted as well. Perhaps some especially tolerant ocean species could endure but the economic and ecological value of the sea would be practically nil.

Is there a solution?

Soviet Central Asia has critical water management problems. Extensive development of irrigation has exhausted surface water resources placing the future of irrigation, the region's economic foundation, in jeopardy. Population continues to grow rapidly which requires an expanded water and food supply as well as increased employment opportunities. Large-scale emigration to labor-short regions in the Russian republic is one possible solution, but most ethnic Central Asians have no desire to relocate where climate, language, and culture are so different from their native lands. The campaign against "cotton at any price," and against centralized planning out of Moscow in general, is especially high in the Uzbek republic (Khazanov, 1990).

Major efforts have been made since 1982, and accelerated under
Gorbachev, to improve irrigation water use: reconstruction of antiquated irrigation systems, water application improvements, automation and computerization, and shifting the crop mix toward lower
water-consuming types. The program has had some success but
much remains to be done. It is planned to establish prices for irrigation water in the near future to promote its careful use. Critics of
irrigation contend that the potential savings from efficiency improvements are enormous – more than enough to meet future regional
needs. Water management specialists see only modest quantities of
water being freed at great cost. Available evidence suggests potential
savings are closer to the thinking of the water management experts
than their opponents.

Available water supplies could be increased by such measures as
greater use of ground water, reuse of irrigation drainage, more regulation of river flow, and the use of water collected in natural basins
and ephemeral streams. All of these, however, have technical problems and economic and environmental costs. The economic structure
of Central Asia could also be refocused: away from water-intensive
irrigation and toward low water use industries (e.g., electronics). This
would require massive increases in regional capital investment and
retraining of the populace, and would not solve the problem of
increasing local food production.

The human-induced drying of the Aral Sea further complicates an
already difficult water management situation. If preventive measures
are not taken, it will shrink to several residual brine lakes in the next
century. The sea's desiccation has caused progressively worsening
environmental, economic, and social problems whose costs, although
difficult to precisely measure, have certainly accumulated into the
tens-of-billions of rubles already. A September 1988 governmental
decree made amelioration of the Aral problem a national priority. The
decree, whereas a step in the right direction, is proving difficult to
implement and will be enormously costly. Furthermore, it will only
preserve a highly saline, much shrunken, and polluted water body.

Until recently, hopes to resolve Central Asia's water problems
rested on future massive water transfers from rivers of Western
Siberia. This project was halted pending further study and validation.
Nevertheless, the Siberian scheme is likely to be resurrected in some
form. There would be a reasonably good chance to solve water problems associated with irrigation in the absence of the Aral Sea's difficulties, or *vice versa*. To adequately deal with both simultaneously

without water importation seems extremely difficult if not impossible.

There are no simple, cheap, or easy answers to Central Asian water management problems. Irrigation expansion based on local water resources, in spite of efficiency measures, will stop in the 1990s. This alone may cause severe social and economic disruption and force large-scale emigration. To provide the inflow the Aral requires to remain ecologically viable, irrigation would have to be substantially reduced, inevitably inducing these consequences. Long-lasting regional enmity toward the central government and communist party with unpredictable political ramifications could result. It would seem a rational Soviet power structure would be unwilling to take such a risk.

An alternative would be diversion of 10 to 15 cu. km annually from Siberian rivers into the Aral Sea, which, along with implementation of some local measures, could preserve the sea near its current level and area while lowering salinity to ecologically tolerable levels – all without any cut-back in irrigation. It could be argued that saving the Aral outweighs the harm to Western Siberia. The Soviet government could, as a condition of the "deal," stipulate that no Siberian water be used for irrigation, encouraging Central Asian water interests to be more efficient, since expansion of irrigation and other water uses would be possible only from water freed by this means.

13   Assessing the environmental impacts of development

In the course of the development of both socialist and capitalist economies, it has all too frequently happened that serious environmental damage has resulted from large landscape-transforming projects. It is all the more regrettable that often such damage might have been prevented by more careful and thoughtful planning of the project. This implies that the probable environmental impacts of proposed large-scale development projects should be thoroughly assessed in advance of the onset of construction. Such studies should not only determine what significant adverse effects might result from the project, but also how such impacts might be mitigated (i.e., reduced in severity), or eliminated altogether.

Environmental analyses of this type have been required on federally funded projects in the United States since passage of the National Environmental Policy Act (NEPA) in 1970, and are called *environmental impact statements* (EIS). One of the most important features of this act is its requirement that alternatives to the proposed project be studied. A number of the individual states likewise require environmental reports on local or state funded projects. Starting in the 1970s, the Soviet Union has also begun to prepare similar types of environmental reviews.

The need for environmental studies and the Soviet response

The goal of environmental impact analysis in the Soviet Union today is to avoid the types of adverse effects that have resulted from many of the large construction projects of the past. Examples of such projects have been given in many of the preceding chapters. They would include the pulp mills which produced the pollution threat to Lake Baikal, and the proposed pipeline to carry the liquid effluents to the Irkut River, which was begun "without waiting for an ecological study" (Sbitnev and Khody, 1987). Also among the cited examples

233

would be the poorly analyzed consequences of the Leningrad flood control dike (chapter 5). And certainly they would include the various hydrotechnical works that have produced the lowered surface levels of the Caspian and Aral Seas.

The level of the Caspian Sea has dropped by several meters from the 1930s to the early 1980s due to water diversions from the Volga and other rivers, and perhaps also from natural climatic causes. This has had harmful effects to fisheries, port facilities, and local microclimates. One proposed response to this problem, the diversion of rivers from northern European Russia into the Volga (alluded to in chapter 12) has not yet been carried out. A second partial solution, the elimination of evaporation from the Kara-Bogaz Gol, was discussed in chapter 11 and serves as another instructive example of the need for advanced environmental analysis. Timely studies of probable environmental consequences might have prevented, or at least alleviated, all of these troublesome projects.

Because of the numerous environmental problems that have occurred, expensive in terms of both money and environment, the earlier practice of promoting construction projects without adequate environmental review has for some time been strongly criticized. Indeed, one historian points out that the first call for environmental impact review in the Soviet Union was made as early as 1929, though not much came of it (Weiner, 1988, p. 92). Chapter 4 noted that the earlier enthusiasm for large hydroelectric reservoirs, which were once routinely approved, is today widely questioned (e.g., Paton, 1986; Podgorodnikov, 1984). By the mid-1970s, the need for improved environmental analysis of major projects had become clear.

The response that developed during the 1980s was the initiation of inter-disciplinary analyses of the effects of at least the largest of new developments on the surrounding environment. Other programs provide comprehensive environmental monitoring in selected industrializing regions throughout the USSR. These efforts are supported by wording in a number of Soviet legal enactments.

The Soviet Union has no single law that mandates environmental impact analyses on all major projects, similar to the National Environmental Policy Act (NEPA) of the United States. NEPA clearly states the circumstances under which formal analyses ("Environmental Impact Statements") must be prepared, and the specific issues that must be evaluated. These evaluations must include a review of alternatives to the proposed project, and means of mitigating adverse impacts. Most importantly, it specifies that the report must be completed, and approved, prior to the start of construction on the project.

In the Soviet Union, which as of 1989 lacked this type of comprehensive legislation, authorization for such environmental studies are found in considerably more generalized wording in other conservation laws, such as the several acts which govern the use of land, water, air, fauna, and other natural resources (Isakov, 1984). These acts contain no explicit requirement for environmental impact studies, but all contain sections that require "conservation," "preservation," and "preventing harmful actions." Though such wording seems to lack an obligatory element, it does suggest that environmental analyses should be carried out routinely as part of the economic planning process (Kolbasov, 1987, pp. 111ff).

The 1980 statute on wildlife management, entitled "Measures for the Protection of the Animal World" contains in Article 21 wording that comes a little closer to mandating environmental impact studies. The wording here requires "the organization of scientific studies aimed at substantiating measures for the protection of the animal world" (Law of the USSR, 1980). However, this article does not specify the types of projects on which such studies will be required, and only alludes inferentially to a need to mitigate adverse effects. Nor does this section specifically require that these studies must be conducted in advance of the start of construction on the project. The new environmental protection law being drafted in 1990 is expected to address these issues more explicitly.

Within the Soviet government, there are numerous agencies that have the technical capability to carry out, or supervise, the preparation of environmental impact analyses. The necessary scientific input can originate from such agencies as the USSR State Committee for Hydrometeorology (Gidromet), which has responsibility for monitoring pollution, from the Chief Directorate for Nature Conservation, Reserves, Forestry, and Hunting, which is responsible for biotic resources and nature preserves, and from a wide variety of scientific-research institutes both within and separate from the USSR Academy of Sciences. The logical reviewing and approval agency for such studies would be the new State Committee on Environmental Protection (Goskompriroda), with input from (among others) the Commission on Environmental Protection and Rational Use of Natural Resources, which is under the Presidium of the USSR Council of Ministers, and the Department of Nature Conservation within the State Planning Agency (Gosplan).

The environmental analyses being carried out in the Soviet Union today are of two basic types. First are regional studies that focus on the environmental effects resulting from one particularly significant large-scale development; in the past these were rarely done prior to the start of construction on the project. The second type are comprehensive monitoring programs to evaluate ongoing environmental changes in a given region from the cumulative effect of all local economic activities. The goal of this type of study is to prepare what are referred to as "Comprehensive Territorial (Regional) Environmental Protection Plans," or TerKSOPs (Vorob'yev et al., 1987). The following sections look briefly at examples of each type.

Environmental studies along the Baikal–Amur Mainline

One of the largest construction projects in Siberia during the 1970s and 1980s was the building of the Baikal–Amur Mainline railroad, usually referred to as the BAM line. This new rail line crosses east Siberia north of the original Trans-Siberian line, and opens up vast new territories for economic development (Shabad and Mote, 1977, chapter 1). It begins at the existing Trans-Siberian rail head at Ust-Kut, runs north of Lake Baikal, and continues eastward to Komsomolsk-na-Amure where an existing line extends it to the Sea of Japan at Sovetskaya Gavan (Figure 13.1).

The environmental ramifications of the project are tremendous, given the severe natural setting in which it is being built (Mote, 1983). They fall into two categories: disruptions caused by the construction of the line itself, and future disruptions resulting from consequent regional development (Sochava, 1977). In response, numerous studies are being carried out in the region of the BAM route, as part of a large comprehensive program entitled "Protecting Siberia's Environment and the Rational Use of Natural Resources" (Luchitski et al., 1983). The objective of these studies is to accumulate baseline environmental data, prior to the advent of major economic transformations, so that long-term changes can be better forecasted and understood (Belov, 1983). One example would be mapping and analyzing the vegetation associations along the BAM route, conducted by the Academy of Sciences' Siberian Institute of Geography (Belov and Krotova, 1981). Another would be the ongoing permafrost research carried out by the Siberian Institute of Permafrost Studies. The environmental forecasting also includes studies on demographic changes, water quality and public health, and recreational potentials. Luchitski reports that the

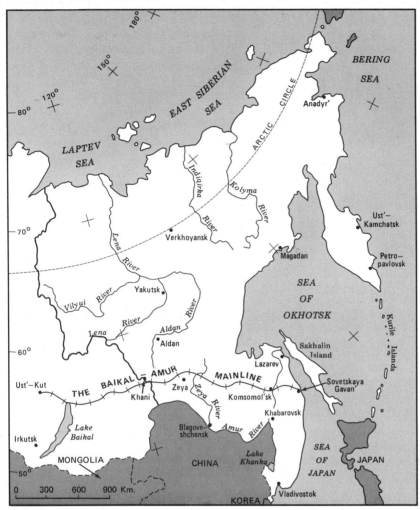

Figure 13.1   Route of the Baikal–Amur Mainline (BAM) *Source: Atlas SSSR, 1983*

entire program on "Protecting Siberia's Environment" (which includes research on KATEK, Lake Baikal, and many other regions) employs nearly 1,400 workers at 32 different research locations, although the number of these specifically associated with the BAM project is not indicated.

It appears that the main focus is on predicting the changes that will be brought about as a result of BAM-induced development. These studies may also include investigations of the local, immediate effects of the work on the BAM line itself, but no references were encountered to any such environmental studies that were completed prior to the start of actual construction on the rail line. Indeed, an article in *Izvestiya* on August 21, 1987, noted that even the detailed engineering designs for the first portions of the project were not approved until 1977, when construction was already underway. Geotechnical studies on roadbed foundations, etc., must have been done at this point, but if studies on the line's disruption of wildlife habitat, for example, were completed prior to the start of construction, they have not been publicized.

The conclusion seems to follow that the emphasis of these environmental studies seems to be much more on forecasting environmental changes than on suggesting mitigation for the adverse consequences of these changes. This apparent shortcoming was noted in an article in *Izvestiya* in 1984:

We became acquainted with a project . . . known as the Territorial Comprehensive Plan for Environmental Protection Along the BAM. Its many volumes provide a comprehensive analysis . . . of the BAM zone and forecast potential pollution and environmental damage. But we failed to find in the plan specific proposals as to how to protect primary topographical features and basic ecosystems . . . and what kind of organizational measures are needed to protect the environment. Yet such proposals are essential. (Druzenko, 1984)

Assuming that the topic of mitigation will ultimately be addressed, the multitude of studies on the effects of the BAM line should constitute an invaluable planning tool. The environmental changes that will be produced by the widespread development of the region will be major, and will justify as many resources as the Soviet government can allocate toward the goal of reducing their impact on the biosphere.

Evaluating the environmental impact of KATEK

In the 1970s, the decision was taken to begin development of the Kansk-Achinsk fuel and energy complex (KATEK), located in the area between the Yenisey River and Lake Baikal in East Siberia. This huge project involves open-pit coal mining and associated thermal power plant construction. The coal-mining portion of the operation could eventually involve removing up to a billion tons of lignite, a widespread but relatively low-grade variety of coal that has a reduced caloric content (Vorob'yev, 1984). The two main centers of mining activity are at Nazarovo and Irsha-Borodino, where earlier small shaft mines have now been replaced with vast open-pit excavations (Figure 13.2).

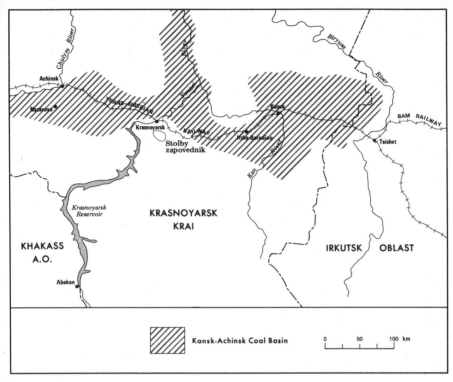

Figure 13.2   The Kansk-Achinsk coal basin *Source: Atlas SSSR,* 1983

Other surface mines were envisioned to be opened up in subsequent phases of the project.

The main use of the lignite which is extracted is to fuel a 1,400 MW power plant located near Nazarovo. Eventually, it was planned that much larger thermal power plants using KATEK coal could transport electricity over high-voltage transmission lines to more westerly portions of the USSR (Shabad and Mote, 1977, p. 50).

A project of this type and magnitude clearly carries significant environmental implications. The potential environmental impacts of the KATEK development have been studied by such bodies as the Institute for Applied Physics in Moscow and the Academy of Sciences' Institute of Geography in Irkutsk. Among the impacts addressed were the effects of the project on climate, air and water quality, forests, and hydrology. The study concluded that the project had considerable potential for environmental damage, and it recommended that ultimate development be limited to just two strip mines and two power stations, instead of four of each as had been originally proposed (Vorob'yev et al., 1987). In an effort to further mitigate potential adverse impacts, the study also recommended construction of a water recycling system, enlarging the size of the cooling reservoirs, and reducing the volume of waste by-products (Izrael' et al., 1981). Other types of environmental studies being carried out by the Siberian Division of the Academy of Sciences in Irkutsk include ongoing monitoring of soils, groundwater, forests, sulfur transport, and atmospheric phenomena (Geografiya, 1984). The results of these and similar studies could result in further modifications to the second phase of the project.

The scope and content of the environmental investigations which have been carried out on the KATEK project is impressive, and in volume and content appears to most resemble a comprehensive environmental impact statement of the type mandated in the United States by NEPA. Further, it appears that a significant scaling back of the project may have resulted from these studies. On the other hand, it is evident that these environmental studies on KATEK were not completed prior to the start of construction on the project, which would be an obligatory requirement under NEPA. Nor can the studies necessarily reduce environmental degradation created by the existing developments, and indeed an article in Pravda on October 17, 1984, complained of severe air pollution resulting from the Kansk-Achinsk project.

The Kursk Model Oblast program

Another type of regional environmental study performed in the Soviet Union is the continuing, comprehensive monitoring of environmental conditions in an already developed region. A term frequently used in the Soviet Union to describe this type of program is "geosystems monitoring." Regional monitoring of this nature has perhaps been most fully developed in Kursk Oblast (Province), in what is termed the "Kursk Model Oblast" project (Kurskaya, 1979; Klyuyev, 1988).

Kursk Oblast is predominantly an agricultural region, located in the rich soils of the European Russian steppe. However, these productive crop lands are interrupted by the huge open-pit iron ore mines of the Kursk Magnetic Anomaly (KMA), which will eventually cover 25,000 ha (Klyuyev, 1988). In addition to the mines, there are also in the region a variety of resource processing and other industrial enterprises, and a large complex of graphite-moderated nuclear power

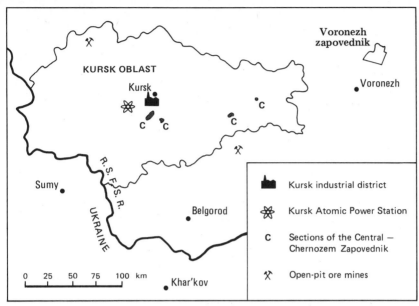

Figure 13.3   Major environmental features of Kursk Oblast

plants, all of which have significant environmental implications (Figure 13.3).

The Kursk Model Oblast project involves numerous research programs on such topics as water quantity and quality, atmospheric contaminants, soil fertility and erosion, the monitoring of pesticides and other toxics, the proper management of overburden materials, and changes in flora and fauna. Some of the most important evaluations concern the effects of the surface mining operations at the Kursk Magnetic Anomaly on the surrounding environment (Figure 11.3). Of primary concern are the effects of the KMA on the water table and topsoils, and consequently on regional agricultural productivity (Zvonkova, 1984). On the other hand, the effects of the agricultural operations themselves (pesticide use, effects on water bodies, etc.) also demand careful attention (Runova and Akhaminov, 1984).

Several of the ongoing research programs on the Kursk region are being conducted by the Academy of Sciences' Institute of Geography at their research station at the Central Chernozem *zapovednik* (Grin, 1984). The author had the opportunity to visit this station in 1983, and was favorably impressed by the variety of monitoring programs underway there (Pryde, 1984). Indeed, this site may be carrying on the most comprehensive monitoring of a steppe ecosystem to be found anywhere in the world today. Additional research is being conducted by personnel from various departments of Moscow State University. One of their studies seeks to predict and evaluate land uses and potential environmental problems in the Kursk Magnetic Anomaly region in the year 2000 (Andreev, 1979).

As its name implies, it is hoped that the Kursk Model Oblast project will be the precursor of similar studies in many other regions of the country, assuming that the sizable funding needed to accomplish this can be made available. As noted at the start of the chapter, the ultimate goal is the preparation of what are termed "Comprehensive Territorial (Regional) Environmental Protection Plans" (Vorob'yev *et al.*, 1987). The KATEK study would be another example. Somewhat similar studies, but on a more limited scale, are being conducted by various research facilities in a number of other regions of the Soviet Union.

Summary of Soviet environmental analysis efforts

The Soviet Union has developed a very broad program for analyzing the probable environmental consequences of major development projects. Although it continues to have no comprehensive legislation that would be analogous to the National Environmental Policy Act (NEPA) in the United States, it does fulfill many of the same objectives by means of research programs carried out by the Academy of Sciences and similar institutions. As a result, the environmental impacts resulting from contemporary major projects such as the Kansk-Achinsk power development or the BAM line are today well analyzed and forecasted, albeit generally only after the project is approved and under construction.

On occasion, they might have even greater import. As noted in chapter 12, the proposals for massive diversions from the Siberian rivers into Central Asia were cancelled in 1987, following the preparation of a large number of environmental studies on the probable consequences of such diversions. It might be possible to fall into a *non sequitur* here, for it is unknown how great a role the environmental studies had in the decision to terminate the project. Most western observers, including the present author, are of the opinion that cost considerations were probably paramount. Nevertheless, the serious environmental implications were widely debated, and could not but have strengthened the hand of those arguing against the project, regardless of their primary motivation.

Large projects, then, are clearly subject to intensive environmental review. It is less clear, however, whether similar environmental review and analysis is required for smaller projects. For example, is the impact on air quality that results from the startup of new but relatively small factories routinely analyzed? Since the air quality impact of one such small factory might be considered insignificant, are the cumulative effects of opening a number of smaller factories in a given region required to be studied prior to the approval of such plans? And if so, what manner of regional (or site-specific) mitigation might be required, and is it in fact carried out in a timely manner?

To carry some of the preceding thoughts a bit further, the foregoing paragraphs relate only to civilian construction projects. Are other types of actions, such as the ploughing of virgin lands for agriculture, or the use of herbicides or insecticides in forestry or farming, subject to formal environmental review? Do military construction projects require environmental analysis? Do construction projects carried out by the USSR in other countries receive detailed environmental analysis? The answer to all these questions is almost certainly "no."

The cases examined in this chapter and the sources cited indicate an emphasis in the Soviet Union on forecasting, or predicting, the effects on the environment of large projects. Although this is highly commendable, the primary objective of environmental analysis is not just to note that the boat has a leak, but also to point out where the repair kit is located and to provide instructions for using it. This implies the identification of possible alternative approaches, and the maximum use of mitigation procedures. However, suggestions for altering projects in order to mitigate adverse impacts have not always been a major component of these studies. Too often the approach has seemed to be: build it first, then study it, and if major problems are observed then scale it back or make changes in the second phase. The Soviet nuclear power plant program epitomizes this syndrome. Even the commendable KATEK study has basically employed this latter approach.

What seems to be needed in the Soviet Union today is a specific national law mandating environmental impact studies, with four main provisions. It should require: (1) that they be carried out on *all* projects that involve environmental deterioration; (2) it should mandate that any reasonable alternatives to the proposed course of action be thoroughly studied; (3) it should require the studies to be completed before work is begun on the project (so that the alternatives can be seriously considered); and (4) it should require mandatory mitigation measures for all significant impacts (as well as long-term monitoring to ensure their effectiveness). The need for environmental analysis to be done prior to construction is starting to be emphasized by Soviet writers (Khorev, 1987). Expanded opportunity for public input would also be helpful. Soviet authorities on environmental law have long urged the passage of some form of a national environmental protection act for the USSR, which would implicitly mandate environmental analysis (Kolbasov, 1987, pp. 123ff). Gorbachev has indicated his support for stronger environmental review as well, having been quoted as saying "all projects must undergo strict scientific analysis to determine possible harm to the environment" (Reuters, Feb. 23, 1989).

A comprehensive nature protection law was being drafted at the time of this writing in 1990 (Zakon . . .). It will address the question of more effective forecasting of the environmental effects of development. A new term that is emerging is "ekspertiza," which is discussed in articles 19–22 of the draft law. Typical usage would be in such easily translated phrases as "projects must undergo *kompleksnaya ekologicheskaya ekspertiza*." It appears to be translatable as "objective review by appropriate experts."

However, such *ekspertiza* can always be ignored, as apparently occurred in the environmental review for the Volga–Chograi irrigation canal, where one scientist maintained that the Ministry of Reclamation virtually ignored the experts' warnings and approved the questionable project anyway, while using the experts' names and the fact that they were consulted as "protection" (Pastukhova, 1988). To counteract such bureaucratic games, an independent agency, which presumedly would be the State Committee on Environmental Protection (Goskompriroda), should have a final review and veto authority.

It is difficult to assess the effectiveness of the environmental impact analysis process in the Soviet Union. Its strong points include large-scale regional monitoring and forecasting, a broad scientific infrastructure, and an apparent commitment to long-term follow-through; weaknesses would seem to include a lack of timeliness, inadequate study of alternatives, and inadequate mitigation provisions. The new law may help to overcome some of these deficiencies.

Over the past two decades, many favorable trends can be identified. Several new nature conservation laws have been enacted, generally with wording that promotes environmental impacts studies. During the same period, the number and sophistication of reports dealing with environmental analysis themes has increased noticeably. Indeed, environmental analysis appears to have been instrumental in scaling back the size of the KATEK project, and may have played an important role in the decision to rescind the plans for the West Siberian river diversions. Without question, the common perception during the Stalin and Khrushchev eras that nature is merely a challenge for the engineering profession (or worse, an annoying obstacle that must be decisively defeated), is now being replaced by an emerging understanding of the critical interrelationship between a healthy and diverse natural environment and the advancement of human well-being.

The long-term beneficial effects on the Soviet biosphere of all these recent efforts cannot be evaluated at present. Although they represent a commendable movement in the direction of improving the quality of the Soviet environment, more needs to be done. The goal of the analytical efforts needs to be expanded from an emphasis on forecasting harmful effects to an emphasis on mitigating or preventing their occurrence. If this can be done on a routine basis throughout the country, a major step will have been taken towards assuring the long-term productivity of the Soviet biosphere and the continuing flow of benefits it can bestow upon the nearly 300 million people who depend upon it.

14 *Glasnost'* and public environmental activism

With the ascent of Mikhail Gorbachev to head the Communist Party of the Soviet Union in 1985, there soon ensued a campaign for more candor in the discussion of Soviet problems. *Glasnost'* quickly became one of the most familiar of all Russian words. Although *glasnost'* has been considered by Gorbachev to be a fundamental component of his effort to restructure and resurrect the country's ailing economic system (*perestroika*), this campaign for a more open debate of issues soon took on a life of its own.

Although environmental issues, at least major ones such as Lake Baikal, had occasionally received considerable public review and debate, the bungled response to the Chernobyl accident, which occurred only a few months after Gorbachev came to power, caused environmental issues to become one of the major focal points of *glasnost'*. Since then, the Soviet public has demanded a much stronger voice in environmental matters. Various aspects of this, such as public concerns about nuclear power, have been discussed in previous chapters. The phenomenon has become so widespread in the Soviet Union at present, however, that a more detailed look at its character and implications is warranted.

Pre-*glasnost'* citizen involvement

The Soviet system has always allowed for a certain amount of public input into environmental issues, although for many years this input was somewhat minimal at the decision-making level. Prior to the *glasnost'* era, citizens could write discreetly worded, though often passionate, letters to newspapers complaining of administrative (but never party) ineptness. Not infrequently, they did. Particularly noteworthy targets were the Lake Baikal controversy (chapter 5), chronic waste in such industries as timber-harvesting (chapter 7), and an effort to spare Leo Tolstoy's country estate at Yasnaya Polyana from the ravages of nearby air pollution.

246

Citizens could also join government-sponsored public conservation groups, such as the All-Russian Society for the Conservation of Nature, discussed below. Even a few local *ad hoc* preservation groups, such as the Leningrad Society for the Protection of Cultural and Historic Monuments, were permitted to operate. Conservation topics have long been included in school curricula (probably to a greater extent than in the United States), and schoolchildren have been encouraged to participate in extracurricular conservation activities.

Leading the public's defense of the Soviet environment were usually scientists and writers, often very well-known ones. The writers in particular, because of their ability to sway public opinion, were very influential in keeping the bureaucrats' feet to the fire. Especially noteworthy in the 1960s were the writings of the popular authors Leonid Leonov (on the timber industry) and Oleg Volkov (on Lake Baikal). In more recent years, other authors who have been outspoken advocates for the environment include Valentin Rasputin (Siberia), Sergey Zalygin (water projects), V. I. Belov (the Volga), and Olzhas Suleimenov (nuclear testing).

The efforts of these writers were greatly needed, for misinformation formerly was widespread. In an earlier review of citizen involvement in conservation issues, it was pointed out that conservation textbooks for many years included some very unfortunate conceptual notions, such as that atmospheric and water resources were "inexhaustible," and that there could be no unwise use of natural resources under socialism (Pryde, 1972, pp. 4–8 and 16–24). A common slogan in the past was "Everything for man, everything for the service of man"; one must wonder what sort of attitudes towards nature this engendered in its listeners.

Another consistent pre-*glasnost'* weakness, noted earlier, in chapter 1, was the tendency in conservation books to cite examples of pollution and natural resources mismanagement only from foreign countries, rarely if ever in the context of the USSR. The widespread environmental damage produced by Stalin's "Great Plan for the Transformation of Nature," dating from the late 1940s era when it was still common to write about the need to "declare war on nature," was sometimes criticized in the 1960s and 1970s, but the idea of the need to systematically transform Soviet nature generally was not. By the late 1980s, however, the massive redesigning of the Soviet landscape had begun to be widely questioned.

Citizen conservation groups have existed in the Soviet Union since its earliest days. However, no nation-wide conservation society has existed since the 1930s; at present such groups are organized and operate in each of the USSR's fifteen constituent republics. The largest is the All-Russian Society for the Conservation of Nature (ARSCN). This Society, which operates throughout the huge Russian Republic, is one of the country's oldest republic-wide conservation organizations, having received its charter from the government on November 29, 1924. The ARSCN's primary duties are educational; it is authorized to conduct public meetings and debates; to organize excursions, laboratories, field stations, museums, libraries, congresses, and university courses; to monitor environmental changes and compliance with conservation laws; and to conduct legal contracts and publish its proceedings (Weiner, 1988, pp. 47, 263). Although its charter was revised and reissued in 1966, these remain its main functions today.

The scope of the ARSCN's activities, and the freedom with which it can conduct them, however, has expanded in recent years; for example, its compliance-with-laws authority gives its members the right to conduct factory inspections, though it must report whatever it finds to other bodies for enforcement. As noted in earlier chapters, however, the ministerial authorities that bear responsibility for the needed remedial actions may be shackled by problems of goals, funding, complacency, or inadequate coordination. The ARSCN does not have the right to initiate litigation (this may change), nor prior to 1989 could its members engage in electioneering.

Other early organizations were the Central Bureau for the Study of Local Lore, composed mainly of scientists, and the All-Ukrainian Society for the Defense of Animals and Plants, both founded in the 1920s (Weiner, 1988). Eventually, similar organizations appeared in all the other Union republics, as well as many organized at more local levels. The first conservation society organized at an institution of higher education was created at Tartu University in Estonia in 1920.

An attempt was made in the 1930s to establish a national conservation organization. At the First All-Union Conservation Congress which was held in January of 1933, there was organized the All-Union Society for the Promotion of the Development and Conservation of Natural Resources. Even its carefully chosen name (embracing the word "development") was not enough to shield many of its members from the effects of Stalin's late-1930s purges, and by World War II it had atrophied.

With the death of Stalin in 1953, the disastrous anti-nature attitudes that had characterized the Stalin era slowly began to give way to the realization that nature merited respect and protection. One strong arena where this manifested itself was in the educational system. The first mandatory conservation course in the country had been initiated at Moscow State University in 1948 (though one must have serious reservations about its content in that era). In more recent decades, conservation elements have been inserted into the curriculum at all educational levels, though, as noted above, they previously included some erroneous ideas. An emphasis on environmental education has been in evidence particularly since the late 1970s, when a number of educational conferences and governmental decrees mandated conservation courses in a wide variety of academic and technical curricula. Programs for training conservation specialists exist at Kazan University, Tartu University, the University of the Urals in Sverdlovsk, and elsewhere. A Laboratory of Nature Conservation Education exists at the USSR Academy of Pedagogical Sciences (Astanin and Blagosklonov, 1983; Koutaissoff, 1987).

As noted above, there are also many opportunities for after-school conservation activities. Schoolchildren have for many decades taken part in such hands-on endeavors as planting trees, erecting nest-boxes for birds, and so forth. Older schoolchildren can participate in "green patrols" to safeguard woodlands or "blue patrols" to help safeguard water bodies. Special days, such as Arbor Day and Bird Day, have become annual events involving practical field exercises. Other extra-curricular environmental activities, such as conservation clubs and school gardens, exist under the leadership of organizations such as the ARSCN.

Efforts to educate the public on environmental issues expanded in the 1970s to include posters and displays in parks, prominent exhibits in museums, high-quality books on environmental topics, and special events. As one example, the Latvian Museum of Nature in the city of Riga contains a large room full of exhibits relating to a wide range of environmental issues. It was one of the best such educational efforts the author has seen in any natural history museum. As regards education, in a 1989 interview Fyodor Morgun, then head of the State Committee for Environmental Protection (Goskompriroda), spoke of a goal for 1995 of having 2,500 new graduates by that date with specific environmental training. In 1990, a popular ecology magazine called *Ekos* began appearing.

However, the general level of citizen environmental awareness still contained some ambivalence at the start of the 1980s. Older people in particular, educated during the Stalin era, tended to view nature as both an attraction and a nuisance, there to be admired and enjoyed, or alternatively overcome or exploited, whichever seems more immediately appropriate. The author has observed well-educated citizens picking flowers in nature reserves despite posted instructions to the contrary, and schoolchildren for decades routinely collected bird nests and eggs as class "nature projects." Slowly, however, the public is becoming better educated not just on the theory, but also on the practice, of environmental conservation.

Glasnost' and the expansion of activism

With the advent of the *glasnost'* era in 1986, hesitancy on the part of the public to voice their opinions on environmental issues very quickly dissipated (Altshuler and Mnatsakanyan, 1990). Instead of merely letters to newspaper editors, street demonstrations became the preferred mode for driving home a point. Suddenly, there were no longer reprisals for this form of civic expression, and the public wasted little time in letting the authorities know exactly where their greatest environmental concerns lay. As Fyodor Morgun put it, "please believe me, the people have awakened" (*Time*, Jan. 2, 1989).

A few examples will show the scope of these concerns. In November of 1987, hundreds of citizens who had organized the Lake Baikal Protection Society (which reportedly has 50,000 members) demonstrated in Irkutsk against the pulp mill wastes that threatened the lake and against the proposed effluent pipeline to the adjacent Irkut River, and helped to stop it (Sbitnev and Khody, 1987). Earlier the same year, 12,000 people marched in protest against emissions from a biochemical plant in Kirishi, a small city near Leningrad. A similar plant proposed near Kazan drew tens of thousands of protests. An *ad hoc* group was formed in Moscow to defeat a housing project in the city's green belt. Many other demonstrations have taken place in Moscow to take advantage of media access. For example, demonstrations have occurred in both Moscow and Semipalatinsk to demand closure of the USSR's nuclear test site in Kazakhstan near the latter city. Anti-nuclear power plant demonstrations have occurred in Krasnodar, Minsk, Yerevan, Odessa, Lithuania, and elsewhere. The location of these and some other protests are shown in Figure 14.1. A surprising number of the above demonstrations have been successful.

An unusual form of nuclear protest occurred in 1989 in the Far East, when fears about its safety prevented the nuclear-powered freighter *Sevmorput* from docking in a number of cities, including the major

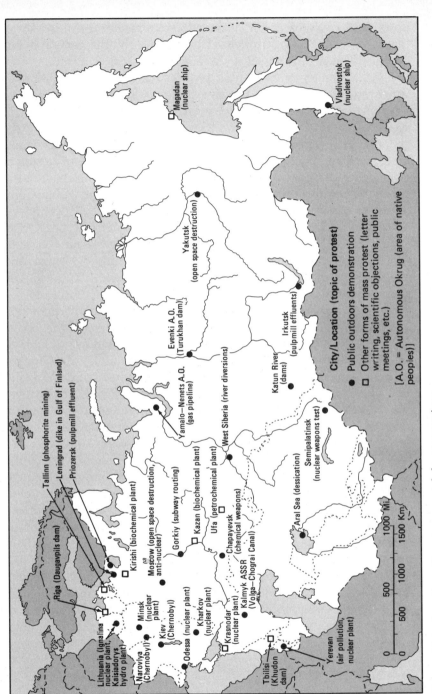

Figure 14.1 Location of environmental demonstrations

ports of Vladivostok and Magadan. Despite assurances of the ship's safety, protests by dockworkers, city Soviets, and the general public successfully kept the vessel at sea (*Sov. Rossia*, Mar. 7, 1989).

A large number of such public protests have focused on proposed water projects. The most massive project was the proposed West Siberian river diversions; a group formed in the Uzbek republic in connection with this issue is the "Save the Aral Sea Committee" (see chapter 12). There have been many other examples. Scientists questioned why the Volga–Chograi irrigation canal was going to be built in Kalmykia if there were to be no additional water diversions into the Volga further north (Pastukhova, 1988). In response, a "Public Committee to Save the Volga" was recently formed by seventy-one influential figures; the group is headed by the writer V. I. Belov (*Sov. Rossia*, Jan. 29, 1989). Indeed, there are now public "rescue" committees formed to help save the Volga, Neva, Ob', Yenisey, Neman, and Dnestr rivers, Lake Ladoga, and the Caspian, Black, Azov, and Aral seas (Wolfson, 1989b).

Leningraders, who have formed at least two local environmental organizations, one called "Green World" and one called the "Delta Group," are protesting the adverse changes to the Gulf of Finland caused by the new dike (see chapter 5). To the south, 30,000 Latvians protested against the proposed Daugavpils hydroelectric dam, and in 1987 plans for it were cancelled (Muiznieks, 1987). People from all over the country have protested against similar dams on the Katun River in West Siberia (chapter 5), forcing further review of the project.

Not all such protest efforts are successful, however. A proposed "In memory of Chernobyl" rally in Kiev in 1988 was disapproved by the city authorities. There are also occasional signs of a pro-development backlash. For example, one article in *Pravda* pointed out the environmental benefits of hydroelectric power plants, and noted that current protests oppose construction of over thirty power plants of various types, which have a total capacity of about 80 million KW that are viewed as needed by the economy (Marchuk, 1989).

The appearance of autonomous environmental groups in the Soviet Union, such as those mentioned above, has been one of the most interesting and significant developments under *glasnost'*. Some of these new groups appear to be almost "establishment." For example, a "Moscow Energy Club" was founded in 1989 to discuss national and world energy problems. An organization calling itself the "Environmental Fund" was created in 1988 to "form and disseminate an environmental world-view," and has links to both the Academy of

Sciences and to Goskompriroda. Indeed, a "Public (advisory) Council" has been formed within Goskompriroda itself (*Pravda*, Jan. 17, 1988). An "Association for the Support of Ecological Initiatives" has been formed in Moscow to assist on both local and global issues (*Environment*, July/Aug. 1989).

Equally new and of great interest are the local "green" groups, which tend to attract activist young people, that have sprung up in all fifteen of the Union republics. In 1990, the Social-Ecological Union of the USSR, perhaps attempting to function as the first nationwide conservation organization, published a list of 331 USSR environmental groups, that included 235 in the Russian Republic, 52 in the Ukraine, and from one to eight in each of the other thirteen republics. Each republic has at least one indigenous green group centered in the republic capital, and multiple green groups exist in such cities as Moscow, Ufa, Odessa, Alma-Ata, Kazan, Gorkiy, Leningrad, Kiev, and elsewhere. Although these organizations are relatively small in terms of membership and lack a high degree of cohesion at present, it is quite probable their numbers and influence will increase with time. As evidence of this, the Latvian "greens" organized themselves in 1990 as the Green Party of Latvia, the first green political party to emerge in the Soviet Union.

Another reason to think that this might happen is that these groups appear to have tacit (and in many cases explicit) governmental approval of their activities. The new State Committee for Environmental Protection (Goskompriroda) is specifically authorized to assist the efforts of local environmental groups (Freeman, 1989). Indeed, the 1988 State of the Soviet Environment report, published by Goskompriroda, devoted a page and a half to describing the new non-governmental ecology organization in the USSR, and listed by name such groups in 14 Soviet cities (Doklad, 1989, pp. 207–8). The Volgograd regional office of Goskompriroda sponsored a tour of citizen environmentalists from the US to inspect local pollution problems in the city and along the Volga.

This approval seems to even include their forging ties with similar green groups abroad, a highly unlikely action prior to the Gorbachev era. For example, the green group in Estonia in 1988 established a significant Soviet precedent by becoming an associate member of the private environmental group "Friends of the Earth." Even more interestingly, the Soviet Peace Committee discussed organizing a subsidiary group, tentatively called "Greenpeace," which might affiliate with the international group of the same name (*Pravda*, June 5,

1987). What makes this so interesting is that the international Greenpeace organization, whose leader was at the meetings where the
above was discussed, is considered one of the most radical of the
western environmental groups and one which in previous years
garnered much worldwide publicity by trying to block the activities of
Soviet whaling ships.

An earlier and much more distinguished group is the International
Physicians for the Prevention of Nuclear War Inc., which has considerable Soviet participation; this group was awarded the Nobel
Peace Prize in 1985. In general, there appears to be considerable
interest in the Soviet Union, as there is in many parts of western
Europe, in linking the anti-war and environmental movements, a
fairly logical coupling.

An unprecedented step was taken in February of 1990, when for the
first time a Soviet environmental organization, the Soviet Association
for Ecology and Peace, intervened in an American environmental
issue. The Association formally requested the US Energy Regulatory
Commission to take steps to protect the sandhill crane during licens-
ing procedures for the Kingsley Dam on the Platte River. The basis for
this request was that a portion of the sandhill crane population breeds
in Siberia. It is very likely that other interventions of this type will
occur in the future.

Nationalism and the issue of local control

The fifteen major nationality groups in the Soviet Union are remarkably diverse (Table 14.1). At the end of the 1980s, a new and
widespread phenomenon among many minority groups in the Soviet
Union was an openly expressed desire for greater national self-
determination (and in some cases, independence). Indeed, by the time
this book is published, the three Baltic republics (at least) may have
declared themselves to be independent countries. One of the individual issues which has fanned the embers of ethnic nationalism is
regional environmental deterioration.

To be sure, there has been no shortage of environmental problems
in the minority republics. The Chernobyl accident would of course
come most quickly to mind, with its resultant demonstrations in not
only the Ukraine, but also in Belorussia, Estonia, and elsewhere.
Other examples, all mentioned earlier in these pages, also exist. These
would include the Aral Sea desiccation and drying up of the Amu-
Darya River, the Kara-Bogaz Gol fiasco, air pollution in Yerevan and

Table 14.1 *Major nationality groups (Union republics) of the USSR*

Republic name	Area (100 mi.²)	% of USSR	Population (1989)	% of USSR	% growth, 1979–89	Language group	Dominant religion
Russian (RSFSR)	6,593	76.2	147,386,000	51	7%	Slavic	E. Orthodox
Estonia	17	0.5	1,573,000	0.6	7	Finno-Ugric	Protestant
Latvia	25	0.3	2,681,000	0.9	6	Baltic	Protestant
Lithuania	25	0.3	3,690,000	1.3	9	Baltic	Roman Catholic
Ukraine	233	2.7	51,704,000	18.0	4	Slavic	E. Orthodox
Belorussia	80	0.9	10,200,000	3.5	7	Slavic	E. Orthodox
Moldavia	13	0.15	4,341,000	1.5	10	Romance	E. Orthodox
Georgia	27	1.9	5,449,000	1.8	9	Caucasus	Georgian Christian
Armenia	12	0.15	3,283,000	1.2	8	Caucasus	Armenian Apostolic
Azerbaidzhan	33	0.4	7,029,000	2.5	17	Turkic	Islam
Kazakh	1,048	12.1	16,538,000	5.8	13	Turkic	Islam
Uzbek	174	2.0	19,906,000	6.9	29	Turkic	Islam
Turkmen	188	1.2	3,534,000	1.3	28	Turkic	Islam
Tadzhik	55	0.6	5,112,000	1.8	34	Persian	Islam
Kirgiz	77	0.9	4,291,000	1.5	22	Turkic	Islam
USSR – Total	8,649		286,717,000		9		

Figure 14.2   "Yesterday Chernobyl, Tomorrow Ignalina"

other non-Russian cities, the Klaipeda oil spill, the Dnestr dam break, the ill-advised draining of the Pripyat marshes, and problems at the Ignalina nuclear plant in Lithuania. Many of these, such as Ignalina, have become the focus of well-organized public opposition campaigns (Figure 14.2). One of the strongest such efforts is that being waged in Kazakhstan against underground nuclear testing; it is being led by Olzhas Suleimenov, a well-known poet and head of the Kazakh Writers Union. This protest, originally entitled the Nevada Movement, changed its name to "Nevada-Semipalatinsk," in reflection of its goal of creating a joint lobbying effort with similar anti-nuclear testing groups in the United States (Figure 14.3).

Prior to *glasnost'*, overt expressions of nationalism were rare, and complaints against environmental degradation were generally voiced on behalf of biosphere protection or economic efficiency, rather than on behalf of "our republic" or "our people." In contemporary Soviet society, the concepts of nationalism and environmental protests have not infrequently become interrelated.

Several causes of this can be identified. One would be the perception that central planners have allowed pollution to grow to an unacceptable level in a given republic. Occasionally, it is implied that

Figure 14.3   Banner of the anti-nuclear testing
"Nevada-Semipalatinsk" movement

Moscow is "dumping" dirty industry into the local area, or that it is
not giving adequate attention to local pollution clean-up.

A second fear is that local natural resources are being exploited
either in an unconscionable manner, or for the benefit of other
regions, with the local ethnic peoples suffering environmental degradation as a result. In some cases this may be more a perception than a
provable situation, but nevertheless some very specific complaints
have been publicized in recent years. One repeated source of such
complaints is Estonia, whose concerns are reviewed below.
Additionally, there have been several allegations of alarming
depletions of wildlife, forests, and other elements of the traditional
resource base of the peoples of the Arctic (Pika and Prokhorov, 1988;

Sizy, 1988). As a result, *Izvestiya* reported on March 10, 1989, that further natural gas development on the Yamal Peninsula in the Yamal-Nenets Autonomous Region of the Arctic would be indefinitely suspended, pending additional studies. A second result has been the formation of an "Association of Small Peoples of the Soviet North," to better protect their cultures and environment; the writer Vladimir Sangi, a native of the Nivkhi people, serves as its first president (*Pravda*, Aug. 9, 1989, p. 2).

Third is a general dissatisfaction with lack of local control over important environmental decisions, a feeling that the environmental fate of minority republics is decided far too much by bureaucrats in Moscow ("Local problems," 1987; Ziegler, 1987, pp. 98ff). There is often a strong belief that the local republic could manage its environmental affairs much better from the republic capital than from the USSR capital.

Finally, not all Union republics are created equal, or so they may feel. There is a perception in some of the smaller ones (e.g., Estonia) that their political clout in Moscow is discouragingly small (Jancar, 1987, p. 198). This probably helps to account for the political alliance and block-voting tendencies of the Baltic republics in the 1989 organizing meetings of the new Soviet congress.

It should be emphasized, though, that not every ethnic area will view a given environmental controversy in the same way. A good example is the 1986 decision to abandon the long-proposed West Siberian river diversions. It was perceived in the Central Asian republics that the main environmental gains from this decision would be in West Siberia (i.e., in the Russian Republic), while the main environmental (and economic) losses would occur primarily in the Kazakh and Uzbek republics (Darst, 1988; Khazanov, 1990; "Local problems," 1987). There is certainly justification for such a feeling, since the sad fate of the Aral Sea has been described as a "water management disaster" and an "ecological calamity" (Kotlyakov, 1988; Micklin, 1988), with no relief immediately in sight for the water problems of Central Asia (Wolfson, 1990).

Also, not only do environmental perceptions sometimes vary spatially, they also vary temporally. That is, the same nuclear power station that might have been welcomed in a local area twenty years ago is viewed with mistrust and fear today. This development, however, probably correlates more strongly to a general increase in environmental awareness than to nationalistic responses.

Associated with this heightened environmental awareness is the universal "nimby" ("not in my back yard") syndrome. This refers to the tendency of people everywhere to agree that toxic wastes, power plants, steel mills, etc., must be located *somewhere*, but just not anywhere near my neighborhood. The Russians are very familiar with this phenomenon, and utilize an almost direct translation, "ne v moyem dvore." The "nimby" syndrome can produce localized protests of very high fervor, which in the USSR often invoke the earlier-mentioned "dumping" charge against Moscow, especially if the individuals involved are an ethnic minority.

The relationship between nationalism and environmentalism is complex, and unwarranted conclusions should be avoided. However, at least two points seem valid. First, to the extent that there is some correlation between the two movements, it is often because in areas where nationalism is already strong, environmental problems become a very convenient focal point for existing energies and organizations. Second, most environmental issues are local, and in cases where the perceived threat is great local residents of all ethnicities will generally band together; nationalistic rallying is much more likely to come into play where the controversy has trans-boundary aspects (Darst, 1988, p. 240).

The vanguard of the activists: Estonia

In the context of public activism, possibly the most interesting regional case study at present is Estonia. This smallest of the fifteen Union republics has taken on a leadership role in terms of both nationalistic and environmental protests. It may well have declared itself independent from the Soviet Union by the time this book is published.

Estonia's environmental consciousness has always been high (Table 14.2). It was the first Soviet republic to enact a comprehensive republic-level conservation law (1957), it created the first national park in the Soviet Union (1971), and in the *glasnost'* era was the first to organize an environmental "green" party (1987). In the late 1980s, the republic was airing both radio and television programs specifically focussed on environmental problems.

Despite the republic's small size, there has been no shortage of environmental issues here. They have included biosphere degradation caused by the extraction of phosphorite and oil shale deposits, pollution from processing and burning the shale, a rumored nuclear waste site near Tallinn, coastal pollution, and regional air quality problems. One survey found that 90 percent of the respondents in the

Table 14.2   *Chronology of Estonian conservation measures*

1910   The Vaika seabird *zakaznik* (preserve) is created
1920   A nature conservation society, the first in Estonia, is established at Tartu University
1935   The first law in independent Estonia is passed concerned with nature conservation
1955   A commission on the conservation of nature is created within the Estonian Academy of Sciences
1957   "Estonian Republic Law on the Conservation of Nature" enacted, the first such law in the USSR
1957   Three new *zapovedniki* (nature reserves) created
1966   A "Society of the Conservation of Nature in the Estonian Republic" is organized
1971   Lahemaa National Park, the first in the USSR, is established
1972   A Commission on Nature Conservation of the Supreme Soviet of Estonia is created
1980   The Estonian Red Book (of endangered species) is published
1988   The Estonian Green Movement is organized
1990s   Independent environmental regulatory agencies and procedures are established

Source: Estonian Natural History Museum, Tallinn

mining center of Kohtla-Jarve concurred that "environmental problems are so critical and significant that immediate remedies must be taken" ("Local problems," 1987, p. 27). This is not a surprising reaction, considering that their children appear to be suffering. According to the newspaper *Sovetskaya Estonia*:

A republic commission appointed to determine why large numbers of children in northeastern Estonia have fallen ill has come to the conclusion that the reason for the children's hair loss and other health abnormalities is, in all likelihood, the pollution of the environment and of daily living conditions in that region . . . In Sillamae, Narva, and Kohtla-Jarve, hair loss has been noted in more than 200 children. We believe that northeastern Estonia should be considered an environmentally critical region, where no more polluting enterprises can be built . . . (Uibu, 1989)

The same article states that the hair-loss phenomenon has also been reported from Novosibirsk, Kemerovo, Mariupol, Makeyevka, and Zaporozhye.

Probably the single greatest object of environmental protest in Estonia is the mining activity (Jancar, 1987, pp. 173f; Taagepera, 1989). Both the phosphorite, used in making fertilizers, and the oil shale, used mainly as a power plant fuel, are surface mining operations that scar thousands of hectares of land. It doesn't help that the mining operations are largely located in a heavily traveled, scenic zone

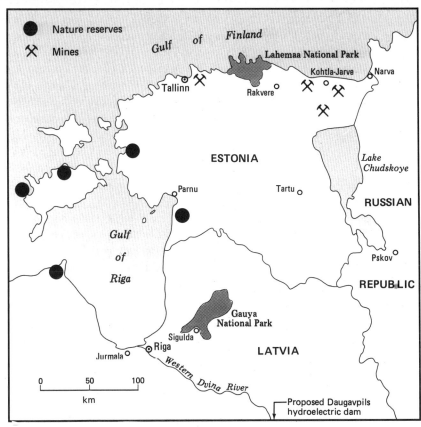

Figure 14.4   Preserved lands and mining centers in Estonia *Source: Atlas SSSR, 1983*

between Tallinn and the republic boundary at Narva, where they are highly visible (Figure 14.4). Estonians also feel their resources are being exploited for the benefit of others, since the fertilizers are of limited use on Estonia's poor soils and much of the electricity produced from the oil shale goes outside the republic.

There is also an interesting complaint that originates from the way industry is organized in the Soviet Union. The ministries that control the mining are All-Union ministries, meaning they are operated nationwide from Moscow, with profits from industry going to the centralized ministry. But new industry in Estonia means that workers, largely Russians, must be imported into Estonia from elsewhere; this is a sensitive point in a republic with a low birth rate that now consists of only about 60 percent Estonians. But the problem is made far worse (in the Estonians' eyes) by the need to pay for the infrastructure to support these newcomers (housing, schools, health facilities, etc.) out of the Estonian SSR budget. All this is in addition to serious questions of pollution of land, air, and water. Thus, it is easy to see why, from the Estonian perspective, current proposals to expand the mining operations are viewed in an entirely negative way and are being adamantly resisted. The Estonians' protests have apparently met with some success; development of a new phosphorite mine near Rakvere has been halted, with some possibility that Estonian phosphorite mining might be phased out in the 1990s (*Soviet Geography*, 1988, pp. 951–3). This will not end the problem, however, as Estonia has most land disturbed by mining *per capita* of any republic (Bond and Piepenburg, 1990, Table 3).

This small republic may also have some of the highest regional air pollution levels of any of the minority republics (Table 14.3). This table quantifies certain aspects of air quality, and compares the results by republic. These data, however, must be viewed as only approximate, as they do not distinguish between types of pollutants involved (some are worse than others), proximity of pollutants to population centers, and other similar qualifying considerations. Nevertheless, the table provides valuable inputs for discussion, and serves as a starting point for further analysis.

An unusual protest took place in Estonia in 1989. The newspaper *Sovetskaya Estonia* on June 3 reported that many Estonian men who had assisted in the clean-up of the Chernobyl disaster were "getting sick, dying." Five were reported to have died, along with at least 227 who require continuing medical treatment. An "Estonian Chernobyl Committee" has been formed; similar committees exist in other republics.

The contemporary environmental protests in Estonia have been led by scientists and writers, reflecting the earlier experience at Lake Baikal and elsewhere, but the general public has been very close behind ("Local problems ...," 1987). The result in 1988 was the formation of the Estonian Green Movement (Devyatkin, 1988; "Some facts ...," 1989). The movement apparently does not maintain a centralized membership roster, but activists in the movement who spoke with me in Tallinn estimated the active membership in 1989 at 4,000–5,000, with perhaps a like number of less active members. Although not defining itself as a political movement (the Popular Front of Estonia plays this role), supporters of the Estonian Green Movement were delegates to the historic 1989 Congress of People's Deputies. Within Estonia, the movement has maintained a close

Table 14.3   *Comparative air pollution indices by Union republic*

Republic	Pollution in tons per capita per year[a]	Pollution in tons per sq. km per year[a]	Cost of pollution control, rubles per capita
Estonia	0.666	41.00	98
Kazakhstan	0.592	4.70[b]	64
RSFSR	0.410	6.00[b]	90
Ukraine	0.360	31.10	66
Turkmen	0.305	4.78[b]	8.2
Belorussia	0.213	10.65	48
Moldavia	0.199	24.00	24
Lithuania	0.191	10.60	35
Azerbaidzhan	0.169	12.60	26
Georgia	0.142	14.20[b]	15
Uzbek	0.117	4.84	19
Latvia	0.115	4.73	33
Armenia	0.100	23.85	3.6
Kirgiz	0.078	1.80	10.4
Tadzhik	0.046	2.97[b]	9.5
USSR average	0.360	6.87[b]	70

[a] Pollution includes the waste products from industry and from transport. The calculations are based on *Narodnoe khoziaistvo SSSR*, 1984, 1985, pp. 403–6; Khachaturov, 1982, p. 91. It is necessary to note that the Soviet statistics do not take into account small industrial sources, the heating of homes and part of the pollution from transport. Therefore the real emissions may be higher by 20–30 percent
[b] In these republics the calculation is not made for the whole territory but for the areas being "economically developed" as suggested by the figures of Khachaturov
Source: "Local problems," 1987, p. 25 (reprinted by permission)

relationship with the republic's political establishment, and virtually all Estonian leaders share the movement's goals.

Most Estonians view attaining full political independence as their ultimate goal. What they would like in the interim is a high degree of *economic* independence from the centralized ministries, with the authority over all industrial decisions (and thus over major sources of pollution) residing firmly in Tallinn, not Moscow ("Some facts," 1989). Indeed, resolutions of "economic independence" have been passed in Estonia (and Latvia and Lithuania as well). In response, the USSR Supreme Soviet in November of 1989 endorsed a proposal for economic independence for the three Baltic republics. Although this brought joy to their area, the idea was earlier opposed by the central planning bodies (who saw their job becoming more difficult), and by other minority republics who questioned (or were envious of?) the preferential treatment (*New York Times*, July 28, 1989).

What the Soviet leadership might find acceptable, as part of *perestroika*, is utilizing the Baltic republics as centers of "new economic thinking" and experimentation, allowing them to be test centers for private enterprise, decentralized planning, and market economy experiments. This could be a way to implement certain goals of *perestroika*, while at the same time giving the Baltic states a measure of the special economic status that they desire. In an environment where the *status quo* is not acceptable to anyone, it might be worth a try. However, with Lithuania declaring itself independent in March of 1990, and with Latvia and Estonia very likely to follow suit, the foregoing concerns might increasingly become a moot point.

Assessing the future of Soviet environmentalism

Soviet environmentalism underwent a fundamental change in the late 1980s. The keystone of this change was that a generalized, but largely silent, public support for environmental improvement was transformed into organized public activism. Citizen spontaneity (in official parlance, "volunteerism") in the past was pointedly discouraged, but for the first time spontaneous, independent citizen organizations such as the Estonian Green Movement have blossomed, seemingly enjoying official tolerance, and at the local level oftentimes even official approval.

Environmental activism is often very high in minority republics (such as Estonia), where it can be linked to other local goals in a common fight against "outside" controls. That is, *glasnost'* often makes it convenient for minority regions to use environmental issues as an "acceptable" (and generally quite justified) context in which to criticize Moscow, Gosplan, the Communist Party of the Soviet Union, *perestroika*, or any other desired target. This marriage of convenience will likely tend to become even stronger in the future, for both causes (nationalism and environmentalism) at present profit from their ties to the other.

For many Soviet environmentalists, however, restructuring – *perestroika* – is central. Just as in the West there are many who believe that permanent environmental improvements are unlikely without some basic changes in the economic/political system, so too there are many in the Soviet Union who believe that fundamental changes in the system are necessary for environmental enhancement. This is no longer a particularly radical notion; it is clear that even Gorbachev, through his *perestroika* campaign, concurs with this to a certain degree. Whether or not such changes are in the process of becoming institutionalized, it is clear that environmental issues have become of fundamental importance to the Soviet public, and will continue to be a significant focal point of public activism for the foreseeable future.

In summary, as the Soviet Union enters the 1990s, environmental consciousness is at an all-time high. This can be attributed to a combination of *glasnost'*, Chernobyl, and improved educational efforts. Environmental education of children appears to be fairly well organized; at least the manifestations of it are widespread both in classrooms and in parks, public exhibits, etc. Hopefully, these efforts to instill an environmental ethic into the next generation will pay future dividends throughout the entire national economy.

However, instilling an environmental awareness into the contemporary bureaucracy is proving not so easy (Wolfson, 1988c). It is always simpler, cheaper (in the short run), and more convenient for ministries to continue to do business in old, familiar ways; change comes very slowly in the Soviet Union. *Perestroika* up to 1990 has been more successful in pointing out problems than in correcting them. The main hope is that the strong environmental component in the wave of change now occurring in the USSR will force the needed restructuring, either through "new thinking" or new personnel, in the staid developmental ministries.

15 International environmental cooperation

Several of the foregoing chapters, particularly those dealing with air pollution and wildlife, have stressed that environmental problems do not stop at international borders. Even if they did, geographic regions of common resource extraction, such as the oceans, would still require international cooperation if their natural bounty is not to be depleted. This is particularly true in the context of mushrooming world population and increasingly sophisticated resource extraction techniques. For all these reasons, international agreements and treaties in the realm of environmental protection have taken on a vital importance in the last thirty years (Kavass and Sprudzs, 1983).

As the world's largest nation in area, second greatest consumer of natural resources, and third biggest in population, the Soviet Union has found it necessary to participate in a great many international agreements concerning environmental problems and natural resources conservation. These include both bi-lateral and multi-lateral agreements, as well as various efforts carried out in conjunction with organizations associated with the United Nations, and even with private groups. This chapter will not attempt to detail all existing accords, but will focus instead on those that are most important, or most typical of Soviet efforts.

Bi-lateral agreements with the United States

Given the economic and military might of the United States and the Soviet Union, the importance of environmental cooperation between the two countries is readily apparent. The first 1988 newsletter of the Conservation Foundation quoted Andrei Sakharov as having stated twenty years earlier that "the salvation of our environment requires that we overcome our divisions and the pressure of temporary, local interests. Otherwise, the Soviet Union will poison the United States with its wastes and *vice versa*." The unfortunately accurate nature of

Table 15.1   *Components of the 1972 US–USSR environmental agreement*

1   Air pollution
2   Water pollution
3   Pollution associated with agricultural production
4   Enhancement of the urban environment
5   Preservation of nature and the organization of preserves
6   Marine pollution
7   Biological and genetic consequences of pollution
8   Influence of environmental changes on climate
9   Earthquake prediction
10 Arctic and subarctic ecological systems
11 Legal and administrative measures for protecting environmental quality

this prophesy has been well documented recently in the context of ozone depletion and other global problems (chapter 2).

Probably the most comprehensive environmental accord between the United States and the Soviet Union was the 1972 Agreement on Cooperation in Environmental Protection, which authorized the exchange of environmental information and specialists in a wide range of scientific fields (Table 15.1). In the ensuing years, hundreds of such exchanges have taken place, including some involving field work in the other country. In 1978 it was my privilege to participate in one such exchange dealing with environmental law and citizen conservation organizations. The pace of US–Soviet cooperative environmental activities declined during the early 1980s, but has been subsequently revitalized, and was specifically mentioned for increased attention in the joint statement that concluded the 1987 Soviet–American summit meeting (Braden, 1990).

The United States and the USSR have also cooperated in a number of other programs aimed at helping particular species; the Siberian crane, which is being captive-bred in both countries to help ensure its survival, serves as one good example. Other wildlife exchanges have involved the reintroduction into Asia of both musk oxen and Prze-walski's horses from North American stocks. Several agreements exist on regulating fishing and crustacean extraction along both oceanic coasts of the United States. A number of these relate to activities in Alaskan waters, with a particular focus on the conservation of various species of salmon and king and tanner crabs. Two related bilateral accords were signed in 1976, one authorizing joint studies on the effects of economic activities on the ecology of the Bering Sea, and one on the conservation of migratory birds (Ginsburgs, 1987).

Figure 15.1   Snow geese on wintering grounds

A particularly suitable symbol of Soviet–American cooperation in the area of wildlife preservation would be the snow goose (*Chen caerulescens*). This hearty resident of the Arctic breeds (in part) in the Soviet Union on Wrangel Island, migrates across Alaska, and winters in the Pacific coast states of the United States (Figure 15.1). If these far-ranging populations of the snow goose are to survive, both countries must cooperate, and indeed a joint research program in the 1970s was able to reverse a serious decline in the Wrangel Island breeding colony. The extension of such cooperative efforts to other areas of transboundary environmental concerns is essential, and should be a major component of both countries' future environmental agenda.

The most widely reported example of Soviet–American cooperation on behalf of wildlife was the October 1988 rescue by a Soviet icebreaker of two gray whales that were trapped in polar ice off Barrow, Alaska. Although the rescue was more symbolic than substantive, the high media exposure it received had a significant educational value concerning the benefits (both biotic and psychological) of Soviet–American cooperation.

In the same area of the arctic, another project is starting to kindle interest. Proposals have been made to create an international "Beringia

National Park," which would take in adjacent portions of the US and the USSR on both sides of the Bering Strait. This narrow (90 km) water body lies between Alaska and the Chukot Peninsula, and once formed the land bridge across which early man travelled from Asia to the western hemisphere. Though the land bridge is now submerged, it is still an important avenue for sea bird, marine mammal, and polar bear migrations, and this feature, combined with its important cultural history, makes it an exciting focus of a cooperative preservation effort between the two countries.

Other areas of cooperation between the US and the USSR include joint research on advancing the promise of nuclear fusion (see chapter 3), and ongoing diplomatic discussions on preventing harmful modifications of the natural environment for military purposes, and reducing the risk of nuclear war. Americans have gone to the Soviet Union to seek out natural predators for such agricultural pests as the Russian wheat aphid, which has recently found its way into the United States, and the leafy spurge, Russian thistle, and a variety of grasshoppers (Associated Press, March 23, 1989).

Significantly, the two countries are starting to use their combined influence to further world-wide conservation efforts. For example, towards the end of 1988 American and Soviet negotiators in Moscow were reported by the Associated Press to have called for a joint enforcement program by the two countries on behalf of curtailing illegal salmon fishing in the north Pacific. While commendable, this is an area that must involve great tact, for many countries are unreceptive to what they may view as patronizing and possibly self-serving advice from even one superpower, never mind two.

A new form of bi-lateral cooperation emerged in the late 1980s when the Soviet Union began working with private conservation organizations on certain specific projects. Probably the most significant of these was the arrangement made in 1987 with the Natural Resources Defense Council, an American non-governmental conservation group, to monitor Soviet underground nuclear tests in Kazakhstan. In 1988, the Soviet Academy of Sciences signed another accord with the same group to conduct research on energy efficiency in buildings and appliances, with the goal of reducing the release of greenhouse effect and ozone-depleting gases. The two organizations are conducting joint energy conservation research at facilities in Minsk, Tallinn, and the Crimea (Watson and Goldstein, 1989). And in the summer of 1989, a group of American and Soviet scientists and environmentalists met in Utah and issued a joint appeal to their respective governments to take more effective action to combat the causes of "greenhouse warming" (see chapter 2).

Agreements to conduct joint research on environment-related projects have also been concluded with private companies, such as Monsanto, and with professional scientific organizations. An important scientific accord was signed in 1988 when representatives from the US and USSR academies of science established a Joint Interacademy Committee on Global Ecology. The Committee plans to investigate and make recommendation on the most pressing global environmental issues of the day. High on their agenda will be the monitoring of global change, including biosphere destruction and many of the atmospheric problems discussed in chapter 2.

In addition to the United States, the USSR has bi-lateral agreements with a great many other countries around the globe.

The United Kingdom signed an agreement with the Soviet Union in 1974 on cooperation and exchanges in the field of environmental protection. This accord was similar to the one signed two years earlier between the US and the USSR, and in subsequent years numerous meetings and projects have been carried out under its terms and provisions.

Bi-lateral accords exist between the Soviet Union and a large number of other countries for purposes of regulating commercial fishing. The USSR has conventions in force to protect marine resources with Norway, Finland, Poland, and several other nations of both Western and Eastern Europe. Similar agreements exist with various Asian countries to regulate the harvest of marine life from the waters of the north Pacific (Ginsburgs, 1987). In all, the Soviet Union has fishing conventions in force with over forty nations, including special reciprocal arrangements with Norway and Japan that permit fishing within the normally protected 200-mile coastal exclusion zone (Kamentsev, 1984).

Procedures to protect endangered species of cranes are in force with Japan, and measures to prevent pollution of the Gulf of Finland have been worked out with the Finnish government. There also exists joint Soviet–Canadian cooperation in the areas of studying the impacts of economic development on Arctic ecosystems, the effects of oil on fish, and the conservation of marine mammals (Astanin and Blagosklonov, 1983).

Multi-lateral agreements

Reflecting the fact that environmental issues often involve a great many nations, multi-lateral accords on environmental protection have become increasingly common. For the Soviet Union, these fall into two main categories, those that primarily involve the East European Council for Mutual Economic Assistance (CMEA) countries, and those that primarily involve western or other non-CMEA countries.

The first significant environmental agreement among the CMEA countries was signed in 1971. In that year, an agreement on joint cooperation on environmental protection went into effect, and to coordinate these efforts, a Joint Council for the Protection of the Environment was established. Given the extent of current environmental problems in Eastern Europe, however, its effectiveness would have to be questioned.

Perhaps the environmental area in which the most effort has been expended by CMEA is water quality, long a major problem in Eastern Europe. The major regional effort in this regard involves the basin of the Tisza River, a major left-bank tributary of the Danube. In the late 1970s an agreement was worked out by Hungary, Czechoslovakia, Yugoslavia, Romania, and the USSR to plan and protect the water resources of this important source of inflow to the Danube, whose drainage basin and waters these countries share.

Similar efforts have been directed towards air quality, and although some factories have been built or retrofitted with abatement equipment, the air quality situation in Eastern Europe remains generally poor. This is particularly true in the more highly industrialized countries such as Poland and Czechoslovakia (Sobelman, 1989; *Environmental Policy Review*, July 1988). The irony here is that the Soviet Union is the downwind recipient of much of the air pollution from East Europe. Perhaps in reflection of this, a special CMEA conference on transboundary pollution was held in 1982. The Soviet Union has ratified the 1979 Convention on Long Range Transboundary Air Pollution (under the UN Economic Commission for Europe) and, as noted in chapter 2, it maintains a series of monitoring stations along its western border to appraise the extent of incoming emissions.

In general, the results of CMEA efforts towards environmental enhancement have not been overly impressive. Ziegler notes that one main reason for this is that CMEA allows member countries to decline to participate in any projects to which they have objections (p. 150). This allows nationalistic CMEA countries, such as Romania, to resist

USSR leadership whenever they choose to do so, and this often leaves CMEA environmental efforts largely confined to research and monitoring.

A number of multi-party agreements have been entered into by the Soviet Union with non-CMEA countries, many of them initiated on behalf of wildlife. One convention of this type that has received much publicity was signed in 1973 with the other Arctic countries of Canada, Norway, Denmark, and the United States for the protection of the polar bear (*Ursus maritimus*). The Soviet Union has established one of its largest nature reserves on Wrangel Island to protect a major concentration of polar bear denning areas. This cooperative program, which also protects the hunting rights of indigenous peoples, has been quite successful, and polar bear numbers are generally considered now to be stable (Uspenskiy, 1989).

A great many international agreements exist aimed at conserving the biotic resources of the earth's regional seas. Multilateral treaties to conserve North Atlantic tuna and salmon were first signed in 1966 and 1982, respectively; with subsequent amendments. The initial international agreements on curtailing whaling were signed in 1956, though not until 1985 did the USSR agree to participate in a ban on taking whales.

One of the more important regional treaties to which the Soviet Union is a party involves the Baltic Sea. The Baltic Sea functions almost like a large saline lake; it is open to the world ocean only via narrow passages through the Danish islands. Water circulation within it is poor, and pollution, originating in both eastern and western nations, pours into it continually. One recent result was that, in the summer of 1988, almost all public beaches along the Baltic coastline were contaminated and a great many were closed to the public. Over a decade earlier in 1974, all the surrounding nations signed the International Convention on the Protection of the Marine Environment of the Baltic Sea in Helsinki, where a permanent secretariate to carry out protection measures for the Baltic was established. To help implement this accord, in 1988 the signatories agreed to reduce by half their discharge of nutrients, organic toxins, and metallic wastes by 1995.

Finland also hosted an international conference on the broader topic of protecting arctic ecosystems in 1989. All eight nations having territory above the Arctic Circle, including the United States and the Soviet Union, participated.

In all, by 1984 the USSR had been party to about twenty regional conventions dealing specifically with environmental topics (IUCN:

"Status . . . ," 1985). Many of these regional agreements were brought into being with the help of various United Nations agencies. Environmental treaties open to all nations (of which the USSR is signatory to about another twenty-five) are summarized in the next section.

The United Nations: USSR participation and universal agreements

The Soviet Union has been active in the affairs of the United Nations ever since its inception. Although many UN actions during its first quarter century were highly political, cooperation through the UN on environmental issues has increased greatly since around 1960. The USSR has regularly participated in these efforts, particularly those of the United Nations Environment Program (UNEP) and UNESCO's Man and the Biosphere Program. As just two examples, the Soviet Union hosted a 1977 UNESCO conference on environmental education in Tbilisi and a 1983 conference on the management of world biosphere reserves. All of Part III of the 1988 State of the Soviet Environment report is devoted to the USSR's international environmental activities, and the first two sections of Part III relate to UN-sponsored organizations and agreements (Doklad . . ., 1989, pp. 223–30).

The United Nations sponsored the creation of the International Union for the Conservation of Nature and Natural Resources (IUCN) in 1948, and for over forty years the Soviet Union has played a leading role in its world-wide conservation efforts. Two Soviet organizations, the Ministry of Agriculture and the All-Russian Society for the Conservation of Nature (see chapter 14), participate in the activities of this organization (Astanin and Blagosklonov, 1983).

The Soviet Union is a party to many UN-sponsored "universal treaties." These are multi-lateral treaties that are open to all nations of the world to sign.

Among the most important of the United Nations' universal treaties are those restricting nuclear weapons. The first significant agreement of this type was the Partial Nuclear Test Ban Treaty of 1963, signed by the Soviet Union, the United States, the United Kingdom, and other nations on August 5 of that year. Without this treaty, cumulative amounts of radioactive isotopes in the earth's atmosphere would by now probably have reached unacceptable levels; near some of the test sites involved they already had. Related treaties designed to reduce risks from nuclear weapons, all signed by these three major powers (and others), include the 1968 non-proliferation treaty, the 1971 treaty banning weapons on the ocean floor, and the 1977 convention prohibiting environmental modification by the military for hostile purposes.

A related area in which there is great need for an effective worldwide treaty involves the banning of chemical weapons of war. The Iraq–Iran conflict, as well as the Soviet action in Afghanistan, brought this issue to the forefront again in the 1980s. A decade earlier, it had centered around the use of defoliants by the United States in Vietnam. The extensive destruction to both the natural environment and to human support systems from the use of chemical weapons can hardly be overstated. At present interest is rising in effecting some kind of mechanism to prevent such destruction in the future. Since such a treaty would have to be multinational, and since enforcement would largely be voluntary, optimism about early or assured world-wide success in this area is difficult. Nevertheless, the US and the USSR agreed bilaterally in 1989 to begin discussions on major reductions in their respective stockpiles of these agents.

The UN has been the focus of efforts to prevent either private or nationalistic exploitation of the resources of the Antarctic continent. The US and USSR, to their credit, have jointly led the effort to discourage individual national claims to portions of that environmentally fragile continent. Several international agreements have been created to protect Antarctica, starting with the prohibition of military bases and nuclear weapons in 1959, and later including treaties protecting its wildlife, limiting mineral exploration, and others. All is not completely well, however. Complaints of over-harvesting of some of the marine resources of adjacent seas persist. More ominously, and encouraged by the US, discussions began in the late 1980s on the possibility of the "international" development of some of Antarctica's resources (particularly coal); the issue will be taken up again in 1990.

Several UN-sponsored treaties have been approved to protect the world ocean and its resources, to which the Soviet Union is a party. These include treaties to prevent oil pollution (1969), to preclude the ocean dumping of wastes (1972 and 1990), and to prevent pollution by substances other than oil (1973). Annex V of the Marpol Convention, (1988), bans dumping plastics at sea, and the 1990) accord bans all industrial waste dumping at sea by 1995. However, compliance with all these provisions is largely voluntary, there is little if any effective enforcement, and the results are often disheartening.

Compared to agreements relating to the oceans, relatively few accords exist that deal with atmospheric pollution. The atmospheric nuclear test ban treaty is perhaps the most important, but for twenty years there were few others. Two more recent agreements, signed by the USSR and discussed earlier in chapter 2, were the 1985 Protocol on the Reduction of Sulfur Emissions, and the 1987 and 1990 treaties banning CFCs so as to protect the ozone layer. Work on a universal treaty was completed in 1989 that would ban the export or "dumping" of hazardous or toxic wastes in other lands; the process of countries signing it had not yet begun at the time of this writing.

A very important wildlife agreement to which the USSR is a party is the Convention Against International Trade in Endangered Species of Fauna and Flora (CITES), which is one of the main defenses against species extinction. Although almost all major nations (including the USA, UK, and USSR) have ratified this convention, for reasons that are not clear almost none of the countries of eastern Europe have signed it.

A related wildlife conservation area in which international agreements need to be expanded and strengthened, both with neighboring countries and countries of destination, concerns the preservation of migrating birds that nest in the Soviet Union, and the protection of their flyways. The need is particularly great in such areas as the Middle East, Southwest Asia, and Indochina.

Summary

The purpose of this chapter has not been to examine in detail all of the environmental treaties to which the Soviet Union adheres, but rather to give an indication of the scope and direction of the USSR's activities in this area. A summary evaluation would be that there are few major international environmental agreements to which the Soviet Union is not a party. However, these conventions vary considerably in their degree of effectiveness, often because of a lack of workable enforcement mechanisms. In closing this chapter, let us consider briefly the efficacy of these international accords.

Multi-national issues concerning environmental protection, like most other topics, are rarely free from subjective national political/ economic considerations. Prior Soviet resistance to complying with international whaling bans, due to long-range economic plans, have already been mentioned (chapter 10). Ziegler points out these nationalistic complexities by citing, first, Soviet complaints about western corporations exporting polluting industries to lesser developed countries where environmental laws might be less stringent; but then also noting that Soviet complaints about "nationalistic ambitions" by some of these same developing nations may really mask Soviet unhappiness about their own fishing boats being excluded from these countries' productive territorial waters (p. 145).

Political by-plays of this sort notwithstanding, these environmental agreements have often had very positive political effects. For example, the 1972 US–USSR environmental cooperation agreement is often cited as one of the cornerstones of "detente," the first post-Vietnam retreat from the most frigid era of the Cold War (Braden, 1990). Similarly, such agreements have often been very positive on a personal basis, and in terms of promoting scientific understanding. Exchanges between Soviet and other scientists have benefited research on environmental problems in a number of countries.

But what have been the results of all these treaties, conventions, and agreements on the actual state of the environment? Can we identify clear environmental improvements stemming from these diplomatic efforts? The results would seem to be mixed.

In some cases, such as the atmospheric test ban treaty, there is little question that the world is a much healthier place than it would have been otherwise. Also, it is generally the case that bi-lateral agreements, which are signed because both parties feel it is to their benefit to do so, are often quite effective. This can particularly be seen in the case of bi-laterial accords to conserve wildlife.

However, certain of the "universal treaties," which often rely on the voluntary compliance of the signatories, may be somewhat slower to produce noticeable improvements. For example, consider the world ocean. Perhaps the plurality of international agreements in effect at present relate in one way or another to the oceans, and without question they have been helpful. Yet today, problems such as oil spills, coastal contamination, over-fishing, ubiquitous plastic flotsam, marine fauna depletion, toxic waste dumping, and closed-sea deterioration (Baltic, Mediterranean, etc.), all attest to the fact that the seas of the world are in serious condition. The efforts to date have not been misdirected, but they certainly have been insufficient, and in some cases, quite ineffective. The current initiatives on behalf of the oceans represent a commendable start, but they are only a beginning. Much stronger efforts will be needed in the near future.

Apart from the oceans, a growing list of other topics also plead for more effective international attention: the greenhouse effect and the ozone hole, tropical forests, acid rain, endangered species, and many more. By the start of the 1990s, the leaders of almost all major countries were expressing their "personal concern" with environmental issues. Hopefully, this is not just campaign rhetoric; the issues are real, serious, and demanding of the most determined international cooperation that can be given to them.

16 The future environment of the Soviet Union

In his 1964 work, *The Meaning of the Twentieth Century*, Kenneth Boulding suggests that the contemporary world is going through a period of very great and rapid transition to a post-developed society, with vast implications for every nation on earth. Later works, such as Alvin Toffler's *Future Shock*, further developed the same theme; but these later works had the advantage of being written in the era of awareness of widespread environmental problems. At the start of the 1990s, one clear meaning of the twentieth century is that human society needs to quickly reform its current attitude towards our planet's environmental support system if the twenty-first century is to have any substantive meaning at all.

The stakes are enormously high. Environmental issues have evolved from local concerns of individual nations into serious global dilemmas that directly or potentially affect everyone. Further, the pace of environmental deterioration seems to be accelerating. This can be clearly seen in such problem areas as species extinction, the destruction of tropical forests, and the absolute numbers of people that are poorly fed and in ill health.

These are difficult challenges that can only be resolved if all countries that have resources available for the struggle put forth a concerted effort towards achieving the needed changes. This must certainly include the Soviet Union. Even if the Soviet leadership wishes to plead that they presently have limited capital resources to devote to world environmental problems, they nonetheless have extensive technical expertise and a demonstrated ability to influence the actions of a great many second and third world nations. These they should employ wisely and fully on behalf of national and world environmental enhancement.

During the early years of environmental awareness, roughly the 1960s and 1970s, attention in the developed countries, including the Soviet Union, United States, and Great Britain, was mainly directed at the most obvious problems: filthy rivers, visible emissions from cars and smokestacks, and so forth. As increasingly sophisticated technologies came into common use, however, pollution grew to be more pervasive, more complex, and in many instances more threatening. As the third millennium draws closer, there would seem to be a compelling need for revised economic and environmental agendas for all nations of the world, and writings on this topic have begun to appear (Bergson and Levine, 1983; Kotlyakov, 1988; Frolov, 1983; *Our Common Future*, 1987).

Perestroika and the Soviet environment

The relationship between Gorbachev's campaign for *perestroika* (political and economic restructuring) and environmental enhancement is an important one. Gorbachev has repeatedly indicated his strong commitment to both issues (Gorbachev, 1987). Indeed, the two are linked, for at present there are many instances where environmental deterioration is adversely affecting economic performance (forestry, agriculture, fisheries, etc.). It may also be that a portion of the reason for the Soviet work force's chronic low productivity involves unhappiness related to environmental issues (particularly those of the workplace). For Gorbachev's economic reforms to be ultimately successful, he must show strong progress in the environmental arena as well.

It is easy to suggest ways in which *perestroika* could produce environmental improvements. Effective reforestation has not been a consistent feature of Soviet timber harvesting in the past; under *perestroika* it should be. Adequate pollution facilities have not always started up simultaneously with new factory openings; under *perestroika* they must. The severe environmental impacts of huge water projects have often gone unstudied prior to project construction; under *perestroika* such impacts should be foreseen and averted; and so forth. The environmental requirements of reconstruction are endless, and involve attitudinal changes throughout the Soviet bureaucracy as much as they do financial inputs.

There is clearly a dilemma here, however, since economic restructuring and environmental enhancement both require considerable commitments of capital. The Soviet Union has acknowledged serious budget deficits in recent years, and capital for all types of projects in the 1990s will be in tight supply. Many laudable goals of *perestroika*, including environmental ones, will find themselves in direct competition for limited budgetary resources. Will new industrial capacity, or promised increases in consumer goods, be sacrificed for additional pollution abatement facilities? In the current era of both *glasnost'* and consumer goods shortages, would the average citizen be willing to accept such a trade-off?

Nor should it be overlooked that *perestroika* itself could have potentially adverse effects on the environment. One of the major goals of restructuring is to increase industrial and agricultural output; but without careful oversight, such increases could easily be accompanied by high environmental costs, as has occurred in the past. As the USSR heads towards both *perestroika* and a "post-industrial" society, the effects of these fundamental transitions on the Soviet environment will need to be continuously monitored.

The linkage between *perestroika* and environmental enhancement can be viewed in another way, in terms of what can be potentially lost. A failure of *perestroika*, with or without the removal of Gorbachev from power, would almost certainly bode ill for the Soviet environment. At present, the most likely alternative to *perestroika* would be a return to the system in effect prior to 1985 (and this system, as embodied in the ministerial bureaucracy, had been little reformed as of the start of the 1990s). This "era of stagnation" system, which *glasnost'* now deems intolerable, produced the vast array of environmental problems with which the Soviet Union is currently trying to cope. In the event of a retreat from *perestroika*, the old institutional failings which produced these problems will likely re-emerge and continue to grind away unchanged (indeed, would be increasingly perceived as being unchangeable).

The environmental consequences of a failure of *perestroika* and a triumph of the old ministerial departmentalism would be enormous. Natural resources would continue to be wastefully utilized. The public health in many specific locations throughout the country would remain in jeopardy. International cooperation on environmental issues, particularly with the United States, would likely be diminished. The average Soviet citizen would not only have to live with continued pollution, but would also suffer the psychological impacts of concluding (at least until proven otherwise) that the system is unreformable, that the bureaucracy is immutable, and that citizen efforts bear little chance of reward.

Hopefully, such a scenario will not come to pass. There is every reason for people both within and without the Soviet Union to work for the success of *glasnost'* and *perestroika*. As of this writing, these reforms appear to offer the best hope for a better environmental future for this large and extremely important portion of the globe.

The Soviet environmental agenda for the 1990s and beyond

The preceding chapters have endeavored to catalog some of the most pressing environmental challenges facing the Soviet Union today. There has been an effort to emphasize the increasing interrelationship between Soviet and world environmental problems, and to show how the Soviet environmental agenda has evolved over the past two decades. This evolution is summarized in Table 16.1. It might be useful to briefly summarize a few of the key issues.

In many people's minds, the most pressing group of environmental problems today involve changes to the earth's atmosphere. Even the most local sources of air pollution inevitably make their own small but important contribution to a general adverse restructuring of atmospheric chemistry. Given the huge size and industrial base of the Soviet Union, the sum of the innumerable point sources of emissions in that country makes a significant collective contribution to such serious worldwide problems as ozone depletion, acid rain, and greenhouse-effect warming. Encouraging steps were taken in the late 1980s in some of these areas, particularly with regard to the ozone problem, but far more needs to be done. Since these problems are intrinsically global in nature, atmospheric deterioration should be near the top of the 1990s environmental agenda of virtually every country on earth.

The same explosion in technology that has produced these atmospheric alterations has engendered a bewildering array of other problems as well. These include more complex forms of pollution of both fresh and marine waters, the broad issue of toxic wastes (including radioactive wastes), and the enormous losses of fish and other biotic resources which these "technological advancements" bring about. The world ocean, like the atmosphere, is increasingly becoming a focal point of these problems, and will require some serious international remedial efforts in the present decade. Many of these problems are now being viewed in a much more serious light, but most are still a very long way from being resolved.

During the 1990s, the Soviet Union will be giving much thought to the future structure of its energy program. Chernobyl produced an inevitable anti-nuclear groundswell in many parts of the country, yet with fossil fuels becoming more expensive and air pollution also of high concern, it is clear that the USSR's nuclear power program can-

Table 16.1   *Changing Soviet environmental emphases*

Environmental topic	1970s and 1980s Agenda				1990 Agenda	2000 Agenda
	Topics in				Topics in 12th FYP[b]	Author's perception[c]
	B.	P.	K.	A.[a]		
Water pollution	X	X	X	X	X	X
Air pollution	X	X	X	X	X	X
Atmospheric change	o					X
Soil erosion	X	X		X	o	o
Land reclamation	o	o	X	o	o	o
Parks and preserves	X	X	X	o	o	o
Forest resources	X	X	o	o		o
Urban greenbelts	o	o		o	o	o
Indoor/work contaminants						X
Minerals, recycling	X	X	o	X	X	X
Fisheries	X	X	o	o		o
Wildlife	X	X	X	X	o	X
Genetic resources				o		X
Lake Baikal		X	X	o		o
Caspian, Aral, Black seas	X	X	X		o	X
Dam, canal projects	o	X	X		X	o
Ocean conservation	o	o		o	o	X
Impact analysis		o			X	X
Pesticides	o	o			o	o
Non-nuclear toxic wastes			o			X
Nuclear energy, wastes	o	o	o	o		X
Nuclear winter						o
World over-population		X				o
Public education	X	o	o	o	o	X
Renewable energy (other than hydroelectricity)		o	o			o

X = major emphasis
o = minor emphasis
FYP = five-year plan
Notes: [a] Sources: B. = Blagosklonov, Inozemtsev, and Tikhomirov, 1967; P. = Pryde, 1972; K. = Komarov, 1980; A. = Astanin and Blagosklonov, 1983
[b] As specifically mentioned in the section "Environmental Protection and the Rational Utilization of Natural Resources" in Draft . . . (1985), pp. 18–19
[c] This column represents the present author's perception in 1989 of year 2000 environmental priorities based on 1989 conditions and trends, and probable actions during the 1990s
Source: Table 16.1 was originally published in the journal *Soviet Geography* (June 1988); permission to re-use the table with minor modifications is gratefully acknowledged

not, and will not, be abandoned. Major changes must be instituted in the design and management of the USSR's commercial nuclear reactors, and though individual proposed locations for new plants will be dropped, others will eventually be selected to replace them. Thus, safe storage of nuclear wastes (including military) will continue to demand a high priority.

But perhaps the need for additional power plants of whatever type in the USSR is exaggerated, and diverts attention from a more basic point, which is energy conservation. One defender of *perestroika* has been quoted as saying that "we don't need any additional energy production, since we spend 1.5 times more energy per unit of GNP than is spent in most western countries" (quoted in Wolfson, 1988b). If the concepts of efficiency and conservation can become key components of Soviet energy planning in the 1990s, there is a good possibility that the need for new power plants, both nuclear and fossil fuel, can be minimized in the USSR, as it has been in the United States.

Indeed, the concepts of natural resource conservation and recycling *must* become an economic way of life in the Soviet Union. Soviet planners should endeavor in the short run to implement a strong energy conservation program, and wherever possible to substitute renewable energy resources, or natural gas, for coal burning power plants. Likewise, much higher levels of solid waste recycling will unquestionably be needed in the twenty-first century. Although the Soviet Union is much less of a "throw-away" society than the United States, both countries can see the desirability of a more effective recycling effort.

The concepts of regional planning, ethnic sensitivities, and environmental management became closely intertwined in the late 1980s in such areas as the Baltic republics, Transcaucasia, and Central Asia. In the latter region, the question of water resources management was in no way resolved by the 1986 decision to leave the Siberian rivers undiverted. Indeed, one could term it an "interim decision," for it seems essential that in the 1990s some form of additional water influxes to the parched Aral Sea basin, and to its burgeoning and underemployed population, be provided.

Virtually all aspects of biotic preservation will require greater attention during the next decade. The problems of poor forestry management and soil erosion have shown scant improvement over the past twenty years, yet both involve resources that will be increasingly essential to the Soviet Union in the future. In both cases, economic restructuring may be able to help, but only if *perestroika* in the forestry and agriculture sectors is instituted with adequate environmental oversight controls.

The worldwide problem of genetic preservation will become more acute in the 1990s. Here, too, the Soviet Union must do more to be a constructive part of the cure. It will predictably create more nature reserves (*zapovedniki*) and other types of protected areas, and hopefully more of these will be in particularly critical habitats in the southern portions of the country. Internal wildlife losses from habitat destruction, pesticide misuse, and poaching must be decreased. But the USSR must also accept a greater leadership role outside its own boundaries, in such areas as the destruction of oceanic species, the protection of migratory birds, and the reduction in slaughter and trade of endangered species.

One form of potential world-wide environmental catastrophe may have been significantly reduced in likelihood during the 1980s, and that is the specter of "nuclear winter." General agreement exists that there would be no winners in a nuclear war, and that long-term devastation would be global (Velikhov, 1985; Harwell and Hutchinson, 1985; *Environment*, June 1988). The agreement by the US and the USSR to destroy their intermediate range missiles is a long overdue first step in the right direction.

The goals of environmental education in the Soviet Union need to be re-focused. The author's perception is that today environmental education in the Soviet school system and in society in general, while certainly capable of being expanded and strengthened, is nonetheless conducted reasonably well. Where the real problem lies is in the abysmal lack of environmental understanding and concern throughout the vast economic bureaucracy. The first task is to make it emphatically clear that *perestroika* implies, among many other things, an encompassing by ministries and enterprises of sound environmental management into their routine agendas. Reference is made to the extensive treatment of this theme in the work by Komarov. The second, more formidable task is to educate a largely environmentally illiterate bureaucracy in the actual ways of accomplishing the needed changes. It can be done only if both carrot and stick incentives are a strong part of the *perestroika* reforms.

Without question there is a need for much more funding for pollution abatement. Where these funds will come from under the current economic situation in the Soviet Union is unfortunately unclear. Reduced emphasis on military expenditures may be part of the answer; eliminating waste and corruption in the economy may be another part. Assuming funds can be made available, there will also need to be more attention paid to the promptness of constructing pollution abatement facilities, to maintaining them, and to strengthening and rigorously enforcing the anti-pollution regulations that have been enacted.

Part of the challenge of effective environmental management is anticipation of problems. Even if significant advances can be realized in all of the above areas, new forms of technologies and pollution will bring about still further problems. Thus, the need for more effective environmental impact analysis, and for associated mitigation measures, will increase. There is at present great difficulty in allocating resources to accomplish today's environmental tasks, much less those that will be identified in the foreseeable future. And even if adequate resources are available (and they are not), extended periods of time may be needed to accomplish some of these goals; problems such as acid rain, for example, cannot be resolved in one or even two five-year plans. Environmental planning must become more long-range; plans need to be formulated now for periods extending well beyond the year 2000.

In summary, the 1990s will demand that the Soviet Union give a much higher level of priority and funding to problems of resource conservation and pollution control. An important part of this will be to force the Soviet bureaucratic apparatus (starting at the Council of Ministers level) to increase its awareness of environmental problems, and to give both biosphere protection and the health of its citizenry a significantly higher priority than they now enjoy. Only in this way can the livability of the Soviet environment and the quality of life of the Soviet peoples not suffer further degradation in the process of restructuring and accelerating the national economy.

Getting there: the search for an environmental road map

What is the Soviet Union presently doing to achieve a more viable environmental future? How does it plan to reach that goal from where it is now?

A variety of efforts have already been undertaken. Previous chapters have noted how, over the years, the Soviet government has issued numerous "special resolutions," or set up special commissions, to try to resolve various environmental problems (Lake Baikal, the Volga River, etc.). Although in all such instances the intentions were sincere, typically, the degree of effectiveness realized ranged anywhere from "somewhat" to "not at all." The need for institutional changes within the ministries that carry out these special decrees has long been evident.

Since the onset of *perestroika* in 1986, many new environmental initiatives have appeared. Perhaps the single most important one was the joint CPSU/Council of Ministers' resolution on the "Fundamental Restructuring of Environmental Protection in the Country" (*Pravda*, Jan. 17, 1988). Its major stipulation was the creation of the new State Committee for Environmental Protection (Goskompriroda), which was discussed in the opening chapter.

Its first head, Fyodor Morgun, made many encouraging statements, but the degree of effectiveness of this new agency is yet to be established (Freeman, 1989; "*Environmental protection . . . ,*" 1989). Morgun's replacement in 1989, Nikolai Vorontsov, is the first non-Communist Party member to head a major governmental agency. Whether this will be a help or a hinderance to the eager biologist-turned-bureaucrat remains to be seen, but it is certainly "new thinking."

One goal of the widespread reorganization of the economy that took place in the late 1980s, with which Goskompriroda will need to concern itself, was to eliminate waste and inefficiency in the national economy. This should in turn help reduce adverse impacts on the environment. Again, it is too early to evaluate results, but it would be unrealistic to expect such changes to appear quickly.

An early effort to improve ecological understanding was a 1987 Academy of Sciences resolution to prepare an interdisciplinary "Program for Biosphere and Ecological Research by the USSR Academy of Sciences Over the Period to 2015." According to one academician, this Program "will occupy a prominent place in the major international geosphere–biosphere program 'Global Changes', currently being formulated . . . for the 1990s" (Kotlyakov, 1988). In the present era of "new thinking," perhaps this effort will lead to more demonstrable results than some of its predecessors.

Another *perestroika* initiative with great potential for environmental reform is the 1987 "Law on State Enterprises." It contains provisions in Article 20 that significantly revise the ways in which factories and other economic enterprises must conduct their environmental business in the future.

Among Article 20's provisions are that: (1) "waste-free technologies" should become the main thrust for future industrial environmental protection; (2) that firms must make comprehensive and rational use of natural resources, and must pay for their use; (3) that mitigation (environmental protection) measures carried out by the factory must fully offset the adverse environmental effects of production; (4) that firms must generally pay for environmental protection, and for wasted natural resources or environmental damage, out of their own budgets; and (5) that in environmental matters, firms fall under the control of local government (i.e., under the local Soviet).

All of the above provisions sound highly laudable, but many of these goals assume the firm has reserves of capital available. Since this has not typically been the case in the past, the question can readily be asked as to where this capital is going to come from.

However, the key provision may prove to be number 5, which takes environmental enforcement away from the ministries and gives it to local authorities. Local Soviets that are tired of polluting industries (and here we must assume that they are not in fact subservient to them) now appear to have a major tool to effect improvements. It will be interesting to see if the phrase "power to the people" might actually become realized under this law. Knowledgeable analysts have been able to point out many reasons why this law may be ineffective (Wolfson, 1988c). Soviet specialists will be watching carefully in the early 1990s for examples of Article 20 being successfully utilized.

If the provisions of this law are implemented faithfully (admittedly a major "if"), it could play a dominant role in correcting not only many harmful environmental practices of the past, but also in reforming the institutional shortcomings which rendered them virtually inevitable. Again, over-optimism should be avoided. But if the provisions of this law are actually carried out and prove effective, then *perestroika* can be said to have achieved an environmental revolution, if not an environmental miracle.

What additional steps might the Soviet Union take to help realize the environmental imperatives of *perestroika*? Throughout the text, the apparent need for a number of new laws has been suggested. Among these are a national law to mandate and govern the preparation of environmental impact analyses on all significant projects. At present, this important task is carried out in a somewhat uncoordinated manner. A law governing the preservation of threatened vegetation is also needed, as is a law to better protect endangered species. These new laws should emphasize mitigation measures and the study of alternative projects. Towards these ends, a draft of a new comprehensive national law on environmental protection was being circulated in 1990 ("Zakon SSSR 'Ob okhrane prirody'").

A longstanding institutional shortcoming that was characteristic of the pre-1986 Soviet Union was a lack of information. In the environmental sphere, this meant a paucity of reliable data on the extent of pollution and other forms of biosphere degradation in the USSR. For example, in such important compilations as UNEP's 1987 *Environmental Data Report* and the Worldwatch Institute's annual *State of the World* report, information on the Soviet Union is highly conspicuous by its near-absence. The same is true for the major study on world air pollution in the October 1989 issue of *Environment* magazine. The Soviet Union needs to quickly become a serious participating member of the global monitoring network, and extend a full measure of *glasnost'* to reporting statistically on the extent of problems within its own borders.

The "rethinking" that has been encouraged under *glasnost'* might be extended into other new areas. A critical examination of the need for new water projects and new nuclear power plants has begun, but what about the contemporary plunge of the country into an automobile dominated society? Have the enormous environmental implications of this commitment been adequately thought through? And what of the really thorny philosophical question of the eventual need for a stabilized economy (that is, one that approaches zero net growth in terms of natural resource usage)? Only rarely has any Soviet official ever spoken out against the Marxist tenets of the primacy of ever-increasing economic growth (see DeBardeleben, 1985, chapter 6).

Nor has the Soviet Union yet developed any coherent policy on the subject of population planning: large families are still encouraged even though certain parts of the country such as Central Asia have an increasing problem of unemployment among young people. Birth control devices are in critically short supply, and as a result abortion rates arc among the highest in the world (*CDSP*, 1987, no. 44, p. 15f.).

Indeed, it has become evident that there is a need for an overall environmental plan, a national environmental ethic, within the Soviet Union. In the past, Soviet environmental programs have almost always been *ad hoc*, single focused, and usually rehabilitatory in nature. Calls have come forth for better regional environmental planning (Khorev, 1987), but what is needed is a comprehensive *national* environmental plan, together with the political determination and necessary funding to successfully bring it to fruition.

Such a plan may be at last taking form. A long-range program for environmental improvement, to encompass not only the 1990–95 five-year plan but also at least a ten-year period beyond, was being formulated in 1989. It is frequently referred to as the "ecology program," and is reportedly several hundred pages long and extremely comprehensive. As of this writing, a draft of this document had not yet appeared. However, at the end of 1989, there did appear the first annual report of Goskompriroda (Doklad, 1989), containing more official data on the state of the Soviet environment than had ever previously appeared under one cover.

As part of Gorbachev's "democratization" efforts, citizens and recognized environmental organizations could be given even more authority than they have at present to correct localized pollution problems. Giving them the right to directly institute legal proceedings might be one approach, and this is reportedly being incorporated into the new Environmental Protection Law that was being drawn up in 1989. *Specialized* citizen organizations are needed; for example, a nationwide "Protectors of the *Zapovedniki*" movement might be created, or a "Friends of Soviet wildlife." The legal position of all "green groups" should be spelled out and guaranteed in legislation, allowing them to be more effective in their efforts; the new "ecology law", if passed, will do this ("Zakon . . ., 1990). A helpful step was empowering Goskompriroda to assist in gaining legal recognition for these groups (Freeman, 1989).

Outside its boundaries, there is an urgent need for the Soviet Union to re-define its role as a potential leader in correcting world environmental problems. Soviet ideological leaders need to stop viewing environmental problems in other countries as strictly a philosophical issue, that is, as solely a consequence of presumed capitalist exploitation, curable by the institution of socialism, as was formerly done. The present economic systems and environmental interdependencies of all the world's nations, including the USSR, are far more complex than this (a fact that would certainly be attested to by all the countries that received fallout from Chernobyl). The reality is that many of the poorest nations lack the necessary resources (of all types) to successfully institute *either* capitalism or socialism.

What is very much needed from the Soviet Union at present is the dedication of more of its own resources to helping third world nations with some of their most difficult problems, such as the destruction of forests or other vital components of their resource base, suboptimal use of the capital they do receive from natural resource exportation, destabilizing rates of population growth, and, in some cases, utterly corrupt political leadership. Soviet aid, and that from other nations, should be on a human scale tailored to village-level problems; giant dams and steel mills are rarely a highest priority (despite some political leaders coveting their "prestige" value), and in some cases have even proven to be counter-productive.

The fact that many developing countries are extremely sensitive to major powers criticizing them on environmental issues (often, it should be pointed out, with good reason) has made it extremely easy for the Soviet Union to garner praise in the third world simply by pursuing a policy of environmental "xeno-myopia." That is, the USSR has tended to refrain from saying anything negative to potentially supportive developing nations, while largely ignoring their critical problems, and letting the United States suffer the adverse consequences of trying to point out where changes are needed. This approach has worked very well for Soviet foreign policy for decades, but if the USSR is really serious about trying to resolve hunger and deprivation in the developing world, such an approach is one they will have to abandon.

Constructive Soviet leadership in world environmental issues would include not only more effective efforts toward ending such problems as desertification and resultant hunger, but also to extending tactful criticism where needed, as for example against those countries that exploit or trade in endangered species, or that still use the deadly miles-long "drift nets" for oceanic fishing which capture everything in their paths. The environmental, health, and food supply problems of the poorer nations require *serious* Soviet commitment to finding viable solutions. These are humane questions, not political ones. The 1990s may be the last decade of opportunity in this regard before even greater ecological disasters descend upon some of the less fortunate third-world countries.

There are some signs that changes along these lines may be occurring. As one example, in April of 1987 Soviet and American scientists issued a joint statement calling for a halt to the destruction of tropical forests. The newspaper *Moscow News* attacked "moral double standards" in a very frank critique of Soviet foreign policy in August of 1989. With regard to foreign environmental policy, the USSR might also consider routinely carrying out environmental impact studies on foreign aid projects that are as detailed as those done internally.

A summing up: the future of the Soviet environment

This chapter opened with the question of the meaning of the twentieth century. Environmentally, its meaning would have to be characterized as self-destructive. It may well be that the twentieth century will be remembered in history as the "Exploitation and environmental entropy century." In that case, the twenty-first century will need to be remembered as the "Stabilization and sustainable symbiosis century." Otherwise, there may not be a twenty-second century, at least not one that we would be proud to have our descendants living in.

The Soviet Union's role in creating this sustainable world environment is great. As the foregoing discussions have noted, in the USSR today *perestroika* and environmental issues are closely related. For this reason, one measure of the effectiveness of *perestroika* will be whether significant improvements are evident during the 1990s in the efficient use of natural resources and in general levels of environmental quality. Put another way, contemporary environmental problems are so serious that if the ensuing decade does not evidence marked improvements in the quality of the Soviet environment, *perestroika* cannot claim to have been a success, no matter what else it may accomplish. The Soviet leadership seems to appreciate this point, and recent policy statements have been encouraging (Gorbachev, 1990). However, lack of funds could thwart such good intentions.

Throughout the world, a recurring obstacle is how slow many political leaders are to realize (or to be willing to admit) that an accelerating environmental crisis faces the entire earth, and that action is needed now. For decades, we have thought of the year 2000 as the last word in long range planning, a far-off time when surely human intelligence would have resolved our environmental enigmas. But the year 2000 is now closer to us than the year 1980. If we really want the third millennium to be an improvement environmentally over the second, we had best replace rhetoric with effective action.

Regrettably, history must deem the 1970s and 1980s as decades of net environmental losses. This is equally true in both the United States and the Soviet Union, where striking parallels exist in the context of environmental problems (Table 16.2). When the present author prepared *Conservation in the Soviet Union* twenty years ago, he concluded it by saying, with regard to environmental amelioration, that "there is still much to be done in both (the USA and USSR), and, as yet, still time to do it." This is no longer the case. Two decades later, time can no longer be viewed as a luxury.

Table 16.2 *Environmental convergence (examples of parallels between environmental problems in the USSR and the USA)*

Issue:	USSR	USA
Diversion of essential water supplies	Aral Sea	Everglades
Nuclear reactor destroyed by negligence	Chernobyl	Three-mile Island
Harmful surface effects from coal mining	Donets Basin	Kentucky, West Virginia
Industrial pollution of a large water body	Lake Ladoga	Lake Erie
Toxic destruction of a water body	Lake Irtyash	Kesterson Reservoir, California
Having to undo a misconceived project	Kara-Bogaz-Gol	Kissimmee River, Florida
High urban air pollution from smelters	Ust-Kamenogorsk	Kellog, Idaho
Expensive project of questionable value	BAM rail line	Tennessee–Tombigbee canal
Montane lake harmed by water diversion	Lake Sevan	Mono Lake, California
Effort to save large migratory bird	Siberian crane	Whooping crane
Large ungulate saved from extinction	European bison	American bison
Large cat now extirpated from the country	Cheetah	Jaguar
Dam threat to popular mountain river	Katun River dam	Stanislaus River, California
River dried up, polluted due to irrigation	Amu-Darya	San Joaquin, California
High urban pollution from steel mills	Magnitogorsk	Gary, Indiana
Severe pollution of harbor of a major city	Leningrad	Boston

A major theme of this work has been that contemporary environmental issues are increasingly global in nature. No nation is an environmental island; indeed, nations never were. This mandates that all countries cooperate fully towards the common goal of resolving the many potential threats to our shared world ecosystem. Fortunately, many recent international events (e.g., the Montreal convention on the ozone layer, the INF treaty, etc.) give cause for hope, but the pace of progress must be accelerated. We must accept the axiom that no nation's environmental problems stand alone, and act accordingly.

The future requires a global understanding that there is only one worldwide environmental agenda on which all nations must work cooperatively. This particularly applies to the Soviet Union, the United States, Japan, the United Kingdom, and all the other developed nations, which must extend a maximum effort toward resolving the current threats to our shared and soiled world ecosystem. These efforts may well involve a considerable amount of environmental *perestroika* within each of these countries' *own* systems. Only by pursuing such a course can we achieve the enlightened leadership, institutional systems, and active citizenry that will be required to pilot spaceship Earth safely into, and through, the twenty-first century.

References

In the citations that follow, *CDSP* refers to *Current Digest of the Soviet Press*; and *SGRT* refers to *Soviet Geography: Review and Translation*.

Abagyan, A. A., *et al.* (1986), "Information on the accident at the Chernobyl nuclear power station and its consequences (prepared for IAEA)," *Soviet Atomic Energy*, 61, pp. 845–68.

Alekseyeva, L., *et al.* (1983), "Specially protected natural areas: reality, problems, and prospects," *Priroda*, 1983, no. 8. pp. 34–43, as translated in *CDSP*, 35 (1983), no. 45, pp. 14–15.

Alferov, Zh. (1983), "Solar energy," *Pravda*, Jan. 31, p. 7.

Alibekov, L. A. (1986), "Narodnyy park Uzbekistana," *Priroda*, no. 1, pp. 41–52.

Altshuler, I. I. and Mnatsakanyan, R. A. (1990), "The changing face of environmentalism in the Soviet Union," *Environment*, 32, no. 2, pp. 4–9 and 26–30.

Andreev, V. V. (1979), "Problems of environmental protection in connection with the formation of the industrial complex of the Kursk Magnetic Anomaly," *SGRT*, 20, pp. 291–6.

Anspaugh, L., Catlin, R. and Goldman, M. (1988), "The global impact of the Chernobyl reactor accident," *Science*, 242, pp. 1513–19.

Anuchin, A. P., *et al.*, eds. (1985), *Lesnaya entsiklopediya*, vol. 1, Moscow, "Sovetskaya entsiklopediya."

Aparisi, R., *et al.* (1980), "Prospects for the development of solar energy in the USSR: production of electric power by thermodynamic methods," *Applied Solar Energy*, 16, no. 6, pp. 1–8.

Aralova, N. S. and Zykov, K. D. (1984), "The role of USSR nature reserves in environmental education," *Conservation, Environment, and Society*, vol. 2, Paris, UNESCO–UNEP, pp. 573–76.

Astanin, L. P. and Blagosklonov, K. N. (1983), *Conservation of nature*, Moscow, Progress Publishers.

Atlas SSSR (1983), Tochenov, V. V. and Markov, V. F. (eds.), Moscow, Glav. Uprav. Geodezii i Kartografii.

Avakyan, A. B. (1987), "Dostoinstva i nedostatki vodokhranilishch," *Priroda*, no. 11, pp. 36–46.

Azimov, S. (1987), "Development and investigations in the area of transformation and use of solar energy in Uzbekistan," *Applied Solar Energy*, 23, no. 5, pp. 1–4.

Bablumyan, S. (1986), "Concerning a preverve," *Izvestiya*, Aug. 11, p. 6, as translated in CDSP, 38 (1986), no. 32, p. 19.

Bakhar', M. F. (1984), *Tayny zhizni lesa*, Minsk, Polymya.

Baklanov, N. (1984), "A shot in the mist," *Izvestiya*, Oct. 29, p. 6, as translated in *CDSP*, 36 (1984), no. 43, p. 26.

Bannikov, A. G. (1968), "Ot zapovednika do prirodnogo parka," *Priroda*, no. 4, pp. 89–97.

(1969), *Zapovedniki Sovetskogo Soyuza*, Moscow, Kolos; 2nd edn, Moscow, Lesnaya promyshlennost, 1977.

Barr, B. and Braden, K. (1988), *The Disappearing Russian Forest*, Totowa, NJ, Rowman and Littlefield.

"Basic provisions of the USSR's long-term energy program" (1984) *Ekon. gazeta*, March, no. 12, pp. 11–14, as translated in *CDSP*, 36 (1984), no. 23, pp. 12–16.

Belgibayev, M. Ye. (1986), "The dust/salt meter – an apparatus for capturing dust and salt in air streams," *Problemy osvoyeniya pustyn*, no. 1, pp. 72–74.

Belousova, L. S. (1967), "Ob organizatsii prirodnykh parkov v Sovetskom Soyuze," in L. K. Shaposhnikov (ed.), *Primechatel'nyye prirodnyye land-shafty SSSR i ikh okhrana*, Moscow, Nauka, pp. 144–54.

Belov, A. V. (1983), "Ecological problems in economic development of the BAM zone" (paper presented at Soviet–American Meeting on the Social–Geographic Aspects of Environmental Change), Irkutsk, Akademia nauk.

Belov, A. V. and Krotova, V. M. (1981), "Geobotanicheskoe rayonirovanie Amurskoi Oblast," *Geografiya i Prirodniye Resursy*, no. 4, pp. 34–43.

Bergholtsas, I. I. (1976), "What should a national park be like?" *Sov. gosudar-stvo i pravo*, no. 1, pp. 72–4, as translated in *CDSP*, 28 (1976), no. 18, p. 8.

Bergson, A. and Levine, H., eds. (1983), *Soviet Economy: Toward the Year 2000*, London, Allen and Unwin.

Blagosklonov, K., Inozemtsev, A., and Tikhomirov, V. (1967), *Okhrana prirody*, Moscow, Izdat. Vysshaya shkola.

Bolin, B. *et al.* (1986), *The Greenhouse Effect, Climatic Change, and Ecosystems*, New York, John Wiley.

Bond, A. (1984), "Air pollution in Norilsk: a Soviet worst case?" *Soviet Geography*, 25, pp. 665–80.

Bond, A. and Piepenburg, K. (1990), "Land reclamation after surface mining in the USSR: economic, political, and legal issues," *Soviet Geography*, May, pp. 332–65.

Boreyko, V. Ye. (1990), "The centennial of the founding of Askaniya-Nova," *Soviet Geography*, 31, pp. 96–107.

Borisenko, Ye and Kondrat'yev, K. (1984), review of S. B. Idso, *Carbon Dioxide: Friend or Foe?*, in *Soviet Geography*, 25, pp. 423–32.

Borisov, V. A. (1968), "Chto zhe takoye natsional'nyy park?," *Priroda*, no. 10, pp. 79–85.

Borodin, A. M., ed. (1985), *Krasnaya kniga SSSR*, Moscow, Lesnaya promyshlennost'.

Borodin, A. M. and Syroyechkovskiy, Ye., eds. (1983), *Zapovedniki SSSR*, Moscow, Lesnaya promyshlennost'.

Boulding, K. E. (1964), *The Meaning of the Twentieth Century*, New York, Harper and Row.

Braden, K. E. (1982), "The geographic distribution of snow leopards in the USSR," *International Pedigree Book of Snow Leopards*, no. 3, pp. 25–39.

(1988), "Environmental issues in Soviet forest management," *Soviet Geography*, 29, pp. 599–607.

(1990), "US–USSR joint committee on protection of the environment," paper presented at the IV World Congress for Soviet and East European Studies, Harrogate, July.

Brasseur, G. (1987), "The endangered ozone layer," *Environment*, 29, no. 1, pp. 6–11 and 39–45.

Browne, M. W. (1983), "A Soviet nuclear boom," *Discover*, 4, no. 12, pp. 18–24.

Budyko, M. I. (1980), *Global Ecology*, Moscow, Progress Publishers.

Budyko, M. I., *et al.* (1990), "Anticipating human modification of global climate," *Soviet Geography*, 31, pp. 11–23.

(1986), "Coming changes in the climate," in I. P. Gerasimov (ed.), *Geographical Prognostication*, Moscow, Progress Publishers, pp. 62–79.

Bugayev, V. (1985), "V krayu zapovednykh trop," *Kazakhstanskaya pravda*, Aug. 30, p. 4.

Campbell, R. W. (1980), *Soviet Energy Technologies*, Bloomington, Indiana University Press.

Cherkasova, M. (1990), "The Katun controversy still on," *Ecos*, 1, no. 1, pp. 62–67.

Chernenko, I. M. (1986), "Problems of managing the water-salt regime of the Aral Sea," *Problemy osvoyeniya pustyn*, no. 1, pp. 3–11.

Chernyshev, Ye. and Barymova, N. (1985), "The role of man in the run-off of dissolved substances," *Soviet Geography*, 26, pp. 159–69.

Cogan, D. G. (1988), *Stones in a Glass House: CFCs and Ozone Depletion*, Washington, DC, IRRC.

"Concerning the first priority measures for improving the use of water resources in the country" (1988), *Vodnyye resursy*, no. 6, pp. 53–55.

Cooper, R. C. (1986), "Petroleum displacement in the Soviet economy: the case of electric power plants," *Soviet Geography*, 27, pp. 377–97.

Darst, R. (1988), "Environmentalism in the USSR: the opposition to the river diversion projects," *Soviet Economy*, 4, pp. 223–52.

DeBardeleben, J. (1985), *The Environment and Marxism–Leninism*, Boulder, Westview Press.

Dergachev, A. (1989), "The sea has spread onto the fields!," *Izvestiya*, Feb. 17, as translated in *CDSP*, 41 (1989), no. 7, pp. 21–22.

Derr, P. *et al.* (1981), "Worker/public protection: the double standard," *Environment*, 23, no. 7, pp. 6–15 and 31–34.

Devyatkin, D. (1988), "An interview with a leader of the Green Movement," *Environment*, 30, no. 10, pp. 13–15.

Dienes, L. and Shabad, T. (1979), *The Soviet Energy System*, New York, John Wiley.

"Doklad: Sostoyaniye prirodnoy sredy v SSSR v 1988 godu" (Report on the state of the Soviet environment for 1988) (1989), Moscow, Gosudarstvennyy Komitet SSSR po Okhrane Prirody (Goskompriroda).

Dollezhal, N. and Koryakin, Yu. (1980), "Nuclear energy: achievements and problems," *Problems of Economics*, 23, no. 2, pp. 3–20.

Dovland, H. (1987), "Monitoring European transboundary air pollution," *Environment*, 29, no. 10, pp. 10–15 and 27–28.

"Draft guidelines for the 12th Five-year Plan" (1985), *Pravda*, Nov. 5, pp. 1–6.

Drizhenko, A. Yu. (1985), *Vosstanovleniye zemel' pri gornykh razrabotkakh*, Moscow, Nedra.

Drucker, G. R. F. and Karpowicz, Z. J. (1989), *Directory of Protected Areas: Eastern Europe and the U.S.S.R.*, Cambridge, IUCN Protected Areas Data Unit.

Druzenko, A. *et al.* (1984), "The law of restitution," *Izvestiya*, Oct. 7, p. 2, as translated in *CDSP*, 36 (1984), no. 40, p. 23.

Dukhovnyy, V. and Razakov, R. (1988), "Aral: looking the truth in the eye," *Melioratsiya i vodnoye khozyaystvo*, no. 9, pp. 27–32.

Dvorov, I. (1986), "Kamchatskoye teplo," *Pravda*, Apr. 8, p. 3.

Eilart, J. (1976), *Man, Ecosystems, and Culture*, Tallinn, Perioodika, pp. 72–7.

Ellis, W. S. (1990), "A Soviet sea lies dying," *National Geographic*, 177, no. 2 (Feb.), pp. 70–93.

"Environmental protection committee, state ecological program – are they enough?" (1989) *Environmental Policy Review*, 3, no. 2 (July), pp. 16–22.

Filipchenko, L. (1989), "Baikal syndrome," *Izvestiya*, May 4, p. 2, as translated in *CDSP*, 41 (1989), no. 18, pp. 28–29.

Flint, V. E., ed. (1984), *A field guide to birds of the USSR*, Princeton, Princeton University Press.

Fokin, A. (1988), "Look back in alarm," *Pravda*, Aug. 4, p. 2, as translated in *CDSP*, 40 (1988), no. 31, p. 22.

Fox, I. K., ed. (1971), *Water Resources Law and Policy in the Soviet Union*, Madison, University of Wisconsin Press.

Freeman, W. E. (1989), "The politics of environmental protection in the USSR: the case of the Soviet EPA," unpublished USIA Research Memorandum, dated Sept. 22.

French, Hilary F. (1990), *Green Revolutions: Environmental Reconstruction in Eastern Europe and the Soviet Union* (Worldwatch Paper 99), Worldwatch Institute, November.

Frolov, I., ed. (1983), *Environmental Protection and Society*, Moscow, USSR Academy of Sciences (Social Sciences Today).

Galaziy, G. I. (1981), "The ecosystem of Lake Baikal and problems of environmental protection," *SGRT*, 22, no. 4, pp. 217–25.

Galeyeva, A. M. and Kurok, M. L., eds. (1986), *Ob okhrane okruzhayushchey sredy*, Moscow, Izdat. politicheskoy literatury.

Gavva, I., Krinitskiy, V., and Yazan, Yu. (1983), "Development of nature reserves and national parks in the USSR," *Parks*, 8, no. 2, pp. 1–3.

Geografiya i prirodnye resursy (1984), no. 1, pp. 30–37 and 159–65; no. 4, pp. 69–97; (1985), no. 4, pp. 118–25 and 130–35; and (1987), no. 3, pp. 10–18.

Gerasimov, I. P., ed. (1971), *Natural Resources of the Soviet Union*, San Francisco, Freeman.

    ed. (1983), *Geography and Ecology*, Moscow, Progress Publishers.

Gerasimov, I. P. and Preobrezhenskiy, V. S. (1983), "National parks as a form of organizing territories for rest and camping," in I. P. Gerasimov, *Geography and Ecology*, Moscow, Progress Publishers, pp. 160–67.

Ginsburgs, G., ed. (1987), *A Calendar of Soviet Treaties, 1974–1980*, Dordreecht, M. Nijhoff.

Gittus, J. H. *et. al.* (1987), *The Chernobyl Accident and Its Consequences*, London, United Kingdom Atomic Energy Authority.

Goldman, M. (1972), *The Spoils of Progress*, Cambridge, The MIT Press.

Gorbachev, M. S. (1987), *Perestroika*, New York, Harper & Row.

    (1990), Speech reprinted in "Two world leaders on global environmental policy," *Environment*, 32, no. 3, pp. 13, 15, 33–35.

Grachev, A. (1985), "Is the water in the Amu-Darya tasty?," *Pravda*, Aug. 13, as translated in *CDSP*, 37 (1985), no. 32, pp. 3–4.

Grin, A. M. (1982), "Goals, tasks, methods and principles of organizing monitoring in the Central Chernozem biosphere reserve," *Isvestiya Akademii nauk SSSR, seriya geog.*, no. 6, pp. 33–40.

    (1984), "The monitoring of geosystems: the case of the Kursk Biosphere Nature Reserve," *Geoforum*, 15, no. 1, pp. 113–22.

Grunbaum, R. (1978), "Alternative energy in the USSR," *Environment*, 20, no. 7, pp. 25–30.

Gunin, P. and Neronov, V. (1985), "Ekologicheskiye printsipy organizatsii zapovednikov v aridnoy zone SSSR," *Problemy osvoyeniya pustyn'*, no. 4, pp. 12–20.

"Hands in the Volga" (1988), *Sov. Rossiya*, Oct. 19, p. 3, as translated in *CDSP*, 40 (1988), no. 43, p. 18.

Harwell, M. and Hutchinson, T. (1985), *Environmental Consequences of Nuclear War* (2 volumes), New York, Wiley.

Hohenemser, C. and Renn, O. (1988), "Chernobyl's other legacy," *Environment*, 30, no. 3, pp. 4–11 and 40–45.

Illesh, A. and Surkov, V. (1985), "The paper phantoms of Tiger Gulch," *Izvestiya*, Jan. 5, p. 3, as translated in *CDSP*, 37 (1985), no. 1, pp. 15–16.

International Union for Conservation of Nature (IUCN) (1985), *Status of Multilateral Treaties in the Field of Environment and Conservation*, Gland, Switzerland, IUCN.

Isachenko, A. G. (1989), "Landshaftovedeniye i zapovednoye delo," *Izvestiya vsesoyusnogo geograf. obshchestva*, no. 4, pp. 277–84.

Isakov, V. (1979), "Rendezvous with nature," *Sovetskaya Rossia*, May 4, p. 4, as translated in *CDSP*, 31 (1979), no. 21, p. 11.

Isakov, Yu. A. (1984), "The protection of nature in the U.S.S.R.: scientific and organizational principles," *Geoforum*, 15, no. 1, p. 93.

Ivchenko, L. (1975), "When resources are unprofitable," *Isvestiya*, Nov. 13, as translated in *CDSP*, 27 (1975), no. 46, pp. 1–2.

Izrael', Iu. A. (1987), "Ecology without cosmetics," *Pravda*, Sept. 7, p. 4, as translated in *CDSP*, 39 (1987), no. 36, pp. 4–6.

(1990), "Chernobyl-1990," Apr. 17, p. 4, as translated in *CDSP*, 42 (1990), no. 18, pp. 8–9.

Izrael', Iu. A. *et al.* (1981), "Basic principles and results of environmental impact studies of the Kansk-Achinsk lignite and power project," *SGRT*, 22, pp. 353–60.

Jackson, W. A. D., ed. (1978), *Soviet Resource Management and the Environment*, Columbus, AAASS.

Jancar, B. (1987), *Environmental management in the Soviet Union and Yugoslavia*, Durham, Duke University Press.

Kaasik, A., and Kask, E. (1983), *Lahemaa rahvuspark (National Park)*, Tallinn, Eesti Raamat.

Kamentsev, V. (1984), "The fish on our table," *Literaturnaya gazeta*, Sept. 12, p. 10, as translated in *CDSP*, 36 (1984), no. 38, pp. 12–13.

Kavass, I. and Sprudzs, A. (1983), *A Guide to United States Treaties in Force* (part 2), Buffalo, W. S. Hein.

Kelly, W., Shaffer, H., and Thompson, J. (1986), *Energy Research and Development in the USSR*, Durham, Duke University Press, 1986.

Kerr, R. A. (1989), "Arctic ozone is poised for a fall," *Science*, 243, pp. 1007–8.

Kes', A. (1978), "Reasons for the level changes of the Aral during the Holocene," *Izvestiya akademii nauk SSSR, seriya geograficheskaya*, no. 4, pp. 8–16.

Khachaturov, T. (1982), *Ekonomika prirodopol'zovaniia*, Moscow.

Khazanov, A. (1990), "The ecological situation and the national issue in Uzbekistan," *Environmental Policy Review*, 4, no. 1, pp. 20–28.

Khorev, B. S. (1987), "On basic directions of environmental policy in the USSR," *Soviet Geography*, 28, pp. 485–89.

Khrenov, Yu. (1985), "Who is indebted to nature," *Izvestiya*, June 8, p. 2, as translated in *CDSP*, 37 (1985), no. 23, p. 22.

Kiselev, V. N. (1984), "Optimal use of land resources in the Belorussian part of the Poles'ye swamps," *Soviet Geography*, 25, pp. 572–84.

Klyuyev, N. N. (1988), "Environmental protection problems and the formation and development of territorial production complexes in developed regions: the KMA," *Soviet Geography*, 29, pp. 839–51.

Knystautas, A. (1987), *The Natural History of the USSR*, New York, McGraw-Hill.

Kolbasov, O. S. (1983), *Ecology: Political Institutions and Legislation*, Moscow, Progress Publishers.

(1987), "Yuridicheskiy mekhanizm obespecheniya ekologicheskoy bezopasnosti," *Nash sovvemennik*, no. 1, pp. 139–43, as summarized in *Environmental Policy Review*, 2 (1988), no. 1, pp. 11–14.

Kolbasov, O. S., Gorokhov, V., and Tranin, A. (1987), "Problemy natsional'nykh parkov Rossii," *Sovetskoye gosudarstvo i pravo*, no. 10, pp. 108–16.

Komarov, B. (1980), *The Destruction of Nature in the Soviet Union*, White Plains, M. E. Sharpe.

Komendar, V. (1987), "Zapovednik ili lesoseka?," *Pravda Ukrainy*, Apr. 8.

Konovalov, B. (1983), "Money to the wind," *Izvestiya*, July 31, p. 2, as translated in *CDSP*, 35 (1983), no. 31, pp. 5, 11, and 20.

Kotlyakov, V. (1988), "Geography and ecological problems," *Soviet Geography*, 29, pp. 569–76.

Kotlyakov, V. and Suprunenko, Yu. (1979), "O sozdanii vysokogornykh lednikovykh natsional'nykh parkov," *Izvestiya Akademii nauk SSSR, seriya geograficheskaya*, no. 5, pp. 25–32.

Koutaissoff, E. (1987), "Survey of Soviet material on environmental problems," ch. 2 in F. Singleton (ed.), *Environmental Problems in the Soviet Union and Eastern Europe*, Boulder, Lynne Rienner.

Kozlov, V. (1984), "When a fish screams," *Izvestiya*, Oct. 4, p. 3, as translated in *CDSP*, 36 (1984), no. 40, pp. 22–23.

Krinitskiy, V. (1981), "Some aspects of territorial organization of biosphere reserves and their role in ecological monitoring," in H. Hemstrom and J. Franklin, eds., *Proceedings, Second US–USSR Symposium on Biosphere Reserves*, Everglades National Park, Sept.

Kukushkin, G. (1986), "Planning the rational utilization of natural resources," *Problems of Economics*, 28, no. 9 (Jan.), pp. 50–61.

Kurskaya Model'naya Oblast' (1979), collection of papers prepared for the Commission on Environmental Problems of the International Geographical Union, Moscow, Academy of Sciences.

Lakhemaa National Park (brochure) (1975), Moscow, Min. Sel'skogo Khozyaistvo SSSR.

Lavrov, S. B. (1990), "Regional and environmental problems of the USSR," *Soviet Geography*, 31, pp. 477–99.

"Law of the U.S.S.R. on air quality" (1980), *Pravda*, June 27, p. 2, as translated in *CDSP*, 32 (1980), no. 28, pp. 9–12.

"Law of the U.S.S.R. on the protection and utilization of the animal world" (1980), *CDSP*, 32, no. 29, pp. 10–14 and 24.

"Law of the U.S.S.R. on the State Enterprise (Association)," *Pravda* (1987), July 1, pp. 1–4, as translated in *CDSP*, 39 (1987), no. 31, pp. 10–17.

Levada, A. (1988), "Will Chernobyl survive?" *Pravda*, Oct. 8, p. 3, as translated in *CDSP*, 40 (1988), no. 40, p. 29.

Lisvenko, N. and Trach, V. (1989), "Gas mask for a city," *Izvestiya*, Oct. 5, p. 3, as translated in *CDSP*, 41 (1989), no. 40, pp. 28–29.

Litvinova, I. (1990), "Why the newspapers didn't come out in Latvia," *Izvestiya*, Jan. 23, p. 3, as translated in *CDSP*, 42 (1990), no. 4, pp. 10–11.

"Local problems (Central Asia without Siberian water; the Estonian phenomenon; Armenians protest)" (1987), *Environmental Policy Review*, 1, pp. 23–32.

Luchitski, I. V., Vorob'ev, V. V., Yermikov, V. D. (1983), "The 'Siberia' comprehensive programme and environmental protection," in V. P. Maksakovski, ed., *The Rational Utilization of Natural Resources and the*

Protection of the Environment, Moscow, Progress Publishers, pp. 197–217.

Lukyanenko, V. (1989), "The drama of water," *Pravda*, Aug. 11, p. 2, as translated in *CDSP*, 41 (1989), no. 32, pp. 29–30.

Lushin, Y. (1988), "Mirages of the Aral," *Ogonyok*, no. 41, pp. 15–19, 33.

Lydolph, P. E. (1977), *Climates of the Soviet Union* (World Survey of Climatology, vol. 7), Amsterdam, Elsevier.

Makeyev, A. (1982), "Sanitation worker or bandit?" *Literatura Rossia*, Apr. 2, p. 22, as translated in *CDSP*, 34 (1982), no. 17, p. 18.

Makhaveyev, N. I. *et. al.* (1984), "A map of erosion-hazardous lands in the non-chernozem zone of the RSFSR," *Soviet Geography*, 25, pp. 390–97.

Maksakovsky, V. P. (1983), *The Rational Utilization of Natural Resources and the Protection of the Environment*, Moscow, Progress Publishers.

Marchuk, A. (1989), "Is a hasty judgement always right?," *Pravda*, Jan. 3, p. 2, as translated in *CDSP*, 41 (1989), no. 1, p. 29; see also *CDSP*, 41 (1989), no. 34, p. 31.

Marochek, V. and Solov'yev, S. (1981), *Pasynki Energetiki*, Moscow, "Znaniye."

Marples, D. R. (1986), *Chernobyl and Nuclear Power in the USSR*, New York, St. Martin's Press.

Medvedev, Zh. (1979), *Nuclear Disaster in the Urals*, New York, Norton.

Mekayev, Yu. (1981), "Nature reserves in the southern European USSR: their condition and future prospects," *Soviet Geography: Review and Translation*, 22, pp. 523–8.

Micklin, P. (1986), "The status of the Soviet Union's North–South water transfer projects before their abandonment in 1985–86," *Soviet Geography*, 27, no. 5, pp. 287–329.

   (1987), "The fate of 'Sibaral': Soviet water politics in the Gorbachev era," *Central Asian Survey*, no. 2, pp. 67–88.

   (1988), "Desiccation of the Aral Sea: A water management disaster in the Soviet Union," *Science*, 241, pp. 1170–76.

   (1989), *The Water Management Crisis in Soviet Central Asia*, final report to the National Council for Soviet and East European Research, contract no. 802:09, Feb. 15.

Micklin, P. and Bond, A. (1988), "Reflections on environmentalism and the river diversion projects," *Soviet Economy*, 4, no. 3, pp. 253–74.

Milanova, Ye. V. and Ryabchikov, A. M. (1979), *Geograficheskiye aspekty okhrana prirody*, Moscow, Mysl'.

Milkov, F. N. (1977), *Prirodnyye zony SSSR*, Moscow, Mysl'.

Mitenkov, F. M. *et al.* (1985), "Reactor plant of the AST-500 heat supply station," *Soviet Atomic Energy*, 58, pp. 339–46.

Morgun, F. T. (1989), "Saving our home," interview in *Argumenty i fakty*, no. 13 (Apr. 1–7), pp. 1–2, as translated in *CDSP*, 41 (1989), no. 18, pp. 29–30.

Mote, V. L. (1978), "Soviet atmospheric resource management," in Jackson, W. A. D. (ed.), *Soviet Resource Management and the Environment*, Columbus, AAASS, pp. 202–14.

(1983), "Environmental constraints to the economic development of Siberia," in Jensen, R., Shabad, T. and Wright, A. (eds.), *Soviet Natural Resources in the World Economy*, Chicago, University of Chicago Press, pp. 17–71.

Mote, V. and ZumBrunnen, C. (1977), "Anthropogenic environmental alteration to the Sea of Azov," *SGRT*, 18, pp. 744–59.

Muiznieks, N. R. (1987), "The Daugavpils hydro station and *glasnost'* in Latvia," *Journal of Baltic Studies*, 18, pp. 63–70.

Myers, N. (1983), *A wealth of wild species*, Boulder, Westview Press.

Narimanov, A. (1981), "Sanitary engineers of the rivers," *Pravda*, July 11, p. 6, as translated in *CDSP*, 33 (1981), no. 28, pp. 18–19.

Narodnoye khozyaystvo SSSR v 1987 g. (1988), Moscow, Financy i statistika.

Nikitin, D. and Novikov, Yu. (1986), *Okruzhayushchaya sreda i chelovek*, 2nd edn, Moscow, Izdat. Vysshaya Shkola.

Nikolayevskiy, A. G. (1985), *Natsional'nyye Parki*, Moscow, Agropromizdat.

Nosov, S. I. *et al.* (1986), *Okhrana zemel'nykh resursov SSSR*, Moscow, Agropromizdat.

Nurberdiyev, M. (1985), "The Kara-Bogaz-Gol's second birth," *Turkmenskaya iskra*, Jan. 10, as translated in *CDSP*, 37 (1985), no. 14, p. 16.

Nuriyev, Z. (1983), "In the interests of present and future generations," *Kommunist*, no. 15, pp. 80–89, as translated in *CDSP*, 35 (1983), no. 44, p. 7.

Okhrana okruzhayushchey sredy i ratsional'noye ispol'zovaniye prirodnykh resursov v SSSR: statisticheskiy sbornik (1989), Moscow, Goskomstat.

Ostachenko, V. F. (1971), "Some safety problems in nuclear plants with water-cooled, water moderated power reactors," *Soviet Atomic Energy*, 30, p. 157.

Our Common Future (Report of the World Commission on Environment and Development) (1987), Oxford, Oxford University Press.

Pallot, J. and Shaw, D. (1981), *Planning in the Soviet Union*, London, Croom-Helm.

Pastukhova, Ye. (1988), "Specialists' arguments that went unheeded," *Izvestya*, Dec. 15, p. 2, as translated in *CDSP*, 40 (1988), no. 50, p. 22.

Paton, B. (1986), "The safety of progress," *Literaturnaya gazeta*, Oct. 29, as translated in *CDSP*, 38 (1986), no. 48, pp. 1–4.

Peterson, C. and Pickard, R. (1988), "Solid waste management practices in the Soviet Union," Appendix F in Peterson, C. W., ed., *Journal: People to People Solid Waste Technology Delegation to the Soviet Union*, report of June 4–18, visit to Moscow, Minsk, and Leningrad.

Peterson, D. J. (1990), "The state of the environment: solid wastes," *Radio Liberty Report on the USSR*, May 11, pp. 11–14.

Petrashkevich, A. (1988), "Salt in the wound," *Literaturnaya gazeta*, Mar. 2, p. 11, as translated in *CDSP*, 40 (1988), no. 10, p. 29.

Petrosyants, A. M. (1987), *Atomnaya nauka i tekhnika SSSR*, Moscow, Energoatomizdat.

Petrov, A., Burdin, N., Kozhukov, N. (1986), *Lesnoy kompleks*, Moscow, Lesnaya promyshlennost'.

Pika, A., and Prokhorov, B. (1988), "Big problems for small peoples,"

Kommunist, no. 16 (Nov.), pp. 76–83, as abstracted in *CDSP*, 40 (1988), no. 47, pp. 22–23.

Pipia, B. (1989), "Cities in the 'Black Book'," *Pravda*, Sept. 1, p. 8, as translated in *CDSP*, 41 (1989), no. 35, p. 27.

Podgorodnikov, M. (1984), "Man-made seas – a time for summing up," *Literaturnaya gazeta*, Oct. 24, p. 11, as translated in *CDSP*, 36 (1984), no. 46, pp. 1–4.

Poletayev, P. (1987), "Responsible for nature," *Pravda*, June 5, p. 2, as translated in *CDSP*, 39 (1987), no. 23, p. 9.

Potter, W. (1990), "Soviet decision making for Chernobyl: an analysis of system performance and policy change," paper presented at the IV World Congress for Soviet and East European Studies, Harrogate, July.

Pralnikov, A. (1985), "Which way is the saiga headed?," *Izvestiya*, Jan. 11, p. 6, as translated in *CDSP*, 37 (1985), no. 2, pp. 17–18.

Precoda, N. (1988), "Leningrad's protective barrier against flooding project," *Soviet Geography*, 29, pp. 725–35.

Presnyakov, A. (1980), "Solar city in the Kopet-Dag," *Pravda*, Sept. 10, p. 6.

"Principles of USSR and Union–republic legislation on land," (1990), *Pravda*, March 7, pp. 2–5, as translated in *CDSP*, 42 (1990), no. 13, pp. 22–28, 36.

Prokhorov, V. (1983), "Can the marmot sleep?," *Pravda*, Oct. 21, p. 3, as translated in *CDSP*, 35 (1983), no. 42 p. 20.

Pryde, P. R. (1967), "The first Soviet national park," *National Parks*, 236, pp. 20–3.

  (1971), "Soviet pesticides," *Environment*, 13, no. 9, pp. 16–24.

  (1972), *Conservation in the Soviet Union*, Cambridge, Cambridge University Press.

  (1978), "Soviet development of non-fossil energy resources," in W. A. D. Jackson (ed.), *Soviet Resource Management and the Environment*, Columbus, AAASS, pp. 148–70.

  (1979), "Geothermal energy development in the Soviet Union," *SGRT*, 20, no. 2, pp. 69–81.

  (1983a), *Nonconventional Energy Resources*, New York, John Wiley.

  (1983b), "The 'decade of the environment' in the U.S.S.R.," *Science*, 220, pp. 274–9.

  (1984a), "Soviet development of solar energy," *Soviet Geography*, 25, pp. 24–33.

  (1984b), "Biosphere reserves in the Soviet Union," *Soviet Geography*, 25, pp. 398–408.

  (1986), "Strategies and problems of wildlife preservation in the USSR," *Biological Conservation*, 36, pp. 351–74.

  (1987), "The distribution of endangered fauna in the USSR," *Biological Conservation*, 42, pp. 19–37.

Rahn, K. A. (1984), "Who's polluting the Arctic?," *Natural History*, 5, pp. 30–38.

Rasputin, V. (1986), "We have only one Baikal," *Izvestiya*, Feb. 17, p. 3, as translated in *CDSP*, 38 (1986), no. 7, pp. 5–7.

Raznoshchik, V. and Lobov, V. (1979), "Attack on solid waste," *Ekon. gazeta*, no. 3, p. 16, as translated in *CDSP*, 31 (1979), no. 4, p. 15.

Reimers, N. (1980), "Which path to take?," *Pravda*, Sept. 8, p. 7, as translated in *CDSP*, 32 (1980), no. 36, p. 18.

(1982), "On a nature reserve path," *Trud*, Dec. 7, p. 3, as translated in *CDSP*, 34 (1982), no. 52, p. 13.

Revkin, A. C. (1988), "Endless summer: living with the greenhouse effect," *Discover*, Oct., pp. 50–61.

Reznikov, A. P. (1989), "Utilizing the solar energy resources of the eastern and high latitude regions of the USSR," *Soviet Geography*, 30, pp. 576–85.

Rosenberg, N. J. (1988), "Greenhouse warming: causes, effects, and control," *Renewable Resources Journal*, Autumn, pp. 4–8.

Roslov, A. (1974), "Reindeer in a trap," *Komsomolskaya pravda*, Jan. 23, p. 2, as translated in *CDSP*, 26 (1974), no. 26, pp. 25–26.

Rozengurt, M. and Herz, J. (1981), "Water, water everywhere, but just so much to drink," *Oceans*, 14, no. 5, pp. 65–67.

Rozov, N. (1971), "Pedagogical description of land resources," in W. A. D. Jackson, ed., *Natural Resources of the Soviet Union: Their Use and Renewal*, San Francisco, Freeman and Co.

Runova, T. G. and Akhaminov, A. D. (1984), "The impact of agriculture on the natural environment," *Soviet Geography*, 25, pp. 733–47.

Ryzhikov, A. I. (1988), "The size of nature reserves and costs of their maintenance," *Soviet Geography*, 29, pp. 917–25.

Ryzhkov, N. (1986), "Basic guidelines for the economic and social development of the USSR in 1986–1990 and in the period to the year 2000," *Pravda*, Mar. 4, as translated in *CDSP*, 38 (1986), no. 12, pp. 1–18.

Sagers, M. J. (1988, 1989), "News notes," *Soviet Geography*, 29, pp. 423–57, and vol. 30, pp. 338–45.

(1989), "Differences in emission of pollutants and inferred air quality among Soviet cities," *Soviet Geography*, 30, pp. 512–20.

Sagers, M. J. and Shabad, T. (1990), *The Soviet Chemical Industry*, Washington, DC, American Chemical Society.

Sapov, V. (1984), "Katun without legend," *Pravda*, July 13, p. 6, as translated in *CDSP*, 36 (1984), no. 28, pp. 17–18.

Sazhin, A. N. (1988), "Regional aspects of dust storms in steppe regions of the East European and West Siberian plains," *Soviet Geography*, 29, pp. 935–45.

Sbitnev, V. and Khody, V. (1987), "Timber cutting near Lake Baikal," *Izvestiya*, Nov. 26, p. 6, as translated in *CDSP*, 39 (1987), no. 47, pp. 8–9.

Scheimer, J. and Borg, I. (1984), "Deep seismic sounding with nuclear explosives in the Soviet Union," *Science*, 226, pp. 787–92.

Schoenfeld, G. (1989), "A dosimeter for every dacha," *Bulletin of the Atomic Scientists*, 45, no. 6, pp. 13–15.

Sebastian, F. (1988), "Municipal waste water treatment," draft paper presented at conference on Soviet Environmental Policies and Practices, University of Glasgow, Mar. 13.

Shabad, T. (1969), *Basic Industrial Resources of the USSR*, New York, Columbia University Press.

(1982), "Vast damage cited in Baltic oil spill," *New York Times*, Jan. 31.

(1984), "Soviet plugs Caspian leak, then restores it," *New York Times*, Nov. 28, p. A–15.

(1986), "Geographic aspects of the Chernobyl nuclear accident," *Soviet Geography*, 27, pp. 504–26.

Shabad, T. and Mote, V. (1977), *Gateway to Siberian Resources (The BAM)*, New York, John Wiley and Sons.

Shalgunov, V. (1984), "Smoke over the Zhiguli hills," *Pravda*, Jan. 5, p. 3, as translated in *CDSP*, 36 (1984), no. 1, p. 11.

Shalybkov, A. and Storchevoy, K. (1988), "Nature preserves: a reference guide," *Soviet Geography*, 29, pp. 589–98.

Shaw, D. J. (1986), "Union of Soviet Socialist Republics," ch. 17 in N. Atricios (ed.), *International Handbook on Land Use Planning*, New York, Greenwood Press.

Shchepotkin, V. (1987), "Nature presents a bill," *Izvestiya*, Sept. 15, p. 2, as translated in *CDSP*, 39 (1987), no. 36, p. 6.

Shimanskiy, M. (1975), "Hunting in foul weather," *Izvestiya*, Feb. 15, p. 5, as translated in *CDSP*, 27 (1975), no. 7, pp. 25–26.

Shipunov, F. (1987), "The Volga's groans," *Sov. Rossiya*, Nov. 18, p. 4, as translated in *CDSP*, 39 (1987), no. 50, pp. 15–17.

Singleton, F., ed. (1976), *Environmental Misuse in the Soviet Union*, New York, Praeger.

Sizy, F. (1988), "The price of Yamal," *Ogonyok*, 46 (Nov.), pp. 20–21, as abstracted in *CDSP*, 40 (1988), no. 47, pp. 23–24.

Sobelman, M. (1989), "New objectives in the area of environmental protection in Poland," *Environmental Policy Review*, no. 1, pp. 22–28.

Sochava, V. B. (1977), "The BAM: problems in applied geography," ch. 8 in Shabad and Mote (1977), *Gateway . . .*, pp. 163–75.

Sokolov, V. (1981), "The biosphere reserve concept in the USSR," *Ambio*, 10, 97–101.

Sokolov, V. and Zykov, K. (1983), "Problems of nature reserve work," *Priroda*, no. 8, pp. 32–33, as translated in *CDSP*, 35 (1983), no. 45, p. 14.

"Some facts from the first year of EGM" (1989), brochure published by the Estonian Green Movement, Tartu.

Soviet Atomic Energy (1983), 54, no. 4 (contains ten review articles regarding Soviet reactors and safety procedures), pp. 243–318.

SSSR v tsifrakh v 1987g. (1988), Moscow, Financy i Statistika, p. 275.

Stebelsky, I. (1987), "Agricultural development and soil degradation in the Soviet Union: policies, patterns, and trends," in F. Singleton (ed.), *Environmental Problems in the Soviet Union and Eastern Europe*, Boulder, CO, Lynne Rienner, ch. 5.

Stewart, J. M. (1987), "The 'lily of birds': the success story of the Siberian white crane," *Oryx*, 21, no. 1, pp. 6–10.

(1990), "The great lake is in great peril," *New Scientist*, no. 1723, pp. 58–62.

Sun, M. (1988), "Environmental awakening in the Soviet Union," *Science*, 241 (Aug. 26), pp. 1033–35.

Taagepera, M. (1989), "The ecological and political problems of phosphorite mining in Estonia," *Journal of Baltic Studies*, 20, pp. 165–74.

Tezikova, T., Obediyentova, G., and Plaksina, T. (1985), "Skhema prirodopol'zovaniya gosudarstvennogo prirodnogo natsional'nogo parka 'Samarskaya luka'," *Izvestiya VGO*, no. 6, pp. 503–11.

Tolmazin, D. (1979), "Black Sea – dead sea?," *New Scientist*, Dec. 6, pp. 767–9.

"Toxicosis of the conscience," (1987), *Literaturnaya gazeta*, Mar. 18, p. 12, as translated in *CDSP*, 39 (1987), no. 19, p. 22.

Tsekov, V. (1974), "In the paper business," *Komsomolskaya pravda*, Jan. 25, as translated, together with follow-up articles, in *CDSP*, 26 (1974), no. 31, pp. 15–16.

Tsymek, A. A. (1975), *Lesoekonomicheskiye raiony SSSR*, Moscow, Lesnaya promyshlennost.

Tunytsya, Yu. Yu. (1987), *Kompleksnoye lesnoye khozyaystvo*, Moscow, Agropromizdat.

Tushinskiy, G. K. (1980), "Perspektivy organizatsii lednikovykh natsional'nykh parkov i bol'shoye Kavkazskoye vysokogornoye kol'tso," *Vestnik Moscov. Univ.*, no. 5, pp. 53–37.

Uibu, J. (1989), "Appeal of the Estonian Republic Ministry of Public Health," *Sovetskaya Estonia*, May 16, p. 3, as translated in *CDSP*, 41 (1989), no. 20, pp. 27–28.

Uspenskiy, S. M., ed. (1989), *Belyy medved'*, Moscow, Min. Sel'khoz.

Valesyan, A. L (1990), "Environmental problems in the Yerevan region," *Soviet Geography*, 31, pp. 573–86.

Vasil'yev, N. F. (1987), "Land reclamation: in the front line of perestroika," *Gidrotekhnika i melioratsiya*, no. 11, pp. 2–10.

Vasil'yev, V. P., ed. (1983), *Okhrana okruzhayushchey sredi pri ispol'zovanii pestitsidov*, Kiev, Urozhay.

Vasil'yev, Yu. *et al.* (1988), "Pyl'nyye buri na yuge Russkoy ravniny," *Izvestiya Akademiya nauk, seriya geograficheskaya*, no. 3, pp. 95–101.

Velikhov, Ye. P., ed. (1985), *The night after: climatic and biological consequences of a nuclear war*, Moscow, Mir Publishers.

Venikov, V. and Putyatin, E. (1981), *Introduction to Energy Technology*, Moscow, Mir Publishers.

Vinokurov, Yu. (1986), "The Katun River's 'cheap' kilowatts," *Pravda*, Dec. 1, as translated in *CDSP*, 38 (1986), no. 48, pp. 4–5.

Volfson (Wolfson), Z. and Rosten, K. (1983), "The world's largest environmental disaster," *Los Angeles Herald Examiner*, Apr. 19.

Volgyes, I. (1974), *Environmental Deterioration in the Soviet Union and Eastern Europe*, New York, Praeger.

Volkov, O. (1977), "The child has outgrown his clothes," *Pravda*, Feb. 22, p. 6, as translated in *CDSP*, 29 (1977), no. 8, pp. 7 and 18.

(1980), "Poacher's path," *Literaturnaya gazeta*, Feb. 6, p. 11, as translated in *CDSP*, 32 (1980), no. 7, pp. 16 and 24.

Vorob'yev, G. I. *et al.* eds. (1979), *Ekonomicheskaya geografiya lesnykh resursov SSSR*, Moscow, Lesnaya promyshlennost'.

Vorob'yev, V. V. (1984), "Problems of protecting the environment in Siberia," *Geoforum*, 15, no. 1, pp. 105–11.

Vorob'yev, V. V. and Martynov, A. (1989), "Protected areas of the Lake Baikal basin," *Soviet Geography*, 30, pp. 359–70.

Vorob'yev, V. V. *et al.* (1987), "Geographic forecasting for optimizing nature management as a basis for the development of the Kansk-Achinsk fuel and energy complex," *Geografiya i prirodnyye resursy*, no. 4, pp. 55–63, as translated in *Soviet Geography*, 29 (1988), pp. 697–707.

Vorontsov, A. and Kharitonova, N. (1977), *Okhrana prirody*, Moscow, Izdat. Lesprom.

Voropayev, G., Blagoverov, B., and Ismayylov, G. (1987), *Ekonomiko-geograficheskiye aspekty formirovaniya territorial'nykh edinits v vodnom khozyaystve strany*, Moscow, Nauka.

Voshchanov, P. and Bushev, A. (1990), "It's easy for children's lives to come to an abrupt end here," *Komsomolskaya pravda*, Apr. 25, p. 2, as translated in *CDSP*, 42 (1990), no. 20, p. 4.

Watson, R. K. and Goldstein, D. B. (1989), "The Natural Resources Defense Council – USSR Academy of Sciences joint program on energy conservation," *Soviet Geography*, 30, pp. 730–5.

Weiner, D. R. (1988), *Models of Nature: Ecology, Conservation, and Cultural Revolution in Soviet Russia*, Bloomington, Indiana University Press.

Whitney, C. R. (1979), "Where caviar comes by the ton," *International wildlife*, 9, no. 6, pp. 4–11.

(1989), "Soviet pollution hurts Finnish wilderness," *New York Times*, May 23, p. B12.

Wolfson, Z. (1988a), "Ecological problems as natural problems: Lake Sevan in Armenia, Lake Baikal and the Volga," *Environmental Policy Review*, 2, no. 2, pp. 3–16.

(1988b), "Non-waste technologies in the USSR," *Environmental Policy Review*, 2, no. 2, pp. 17–22.

(1988c), "Perestroika and glasnost' in environmental policy," *Environmental Policy Review*, 2, no. 1, pp. 1–25.

(1989a), "Dangerous levels of pesticides and other chemicals in food," *Environmental Policy Review*, 3, no. 1, pp. 7–11.

(1989b), "The fatherland is in ecological danger: Chernobyl survey," *Environmental Policy Review*, 3, no. 2, pp. 9–15.

(1990), "Central Asian environment: a dead end," *Environmental Policy Review*, 4, no. 1, pp. 29–46.

Wynne, B. (1989), "Sheepfarming after Chernobyl," *Environment*, 31, no. 2, pp. 10–15 and 33–39.

Yablokov, A. V. (1988), "Pesticides, ecology, and agriculture," *Kommunist*, no. 15, pp. 34–42, abstracted in *CDSP*, 41 (1989), no. 2, pp. 20, 32.

(1990), "The current state of the Soviet environment," *Environmental Policy Review*, 4, no. 1, pp. 1–19.

Yeliseyev, N. (1966), "Ne tol'ko pod Yaroslavlem . . .," *Komsomol'skaya pravda*, Dec. 23, p. 4.

Yeliseyeva, V. I. (1976), "Raznogodichnaya dinamika naseleniya ptits v dubravakh Tsentral'no-chernozemnogo zapovednika," in A. M. Grin (ed.), *Biota osnovnykh geosistem tsentral'noy lesostepi*, Moscow, Akademiya nauk SSSR.

Zakharko, V. (1983), "Sept. 15 dam break threatened water supply," *Izvestiya*, Oct. 27, p. 6, as translated in *CDSP*, 35 (1983), no. 42, pp. 1–3.

"Zakon SSSR 'Ob okhrane prirody' (proyekt)" (Law of the USSR 'On the protection of nature'), draft being circulated for review in 1990.

"Zakon SSSR ob osobo okhranyayemykh prirodnykh territoriyakh (proyekt)" (Law of the USSR on specially protected natural areas), draft being circulated for review in 1990.

Zalogin, N. G. *et al.* (1979), *Energetika i okhrana okruzhayushchey sredy*, Moscow, Energiya.

Zalygin, S. (1984), "Moving water, standing water," *Izvestiya*, Oct. 20, p. 2, as translated in *CDSP*, 36 (1984), no. 46, pp. 4–5.

"Zapovedniki, zapovedno-okhotnichi khozyaistva i natsionalnyye parki SSSR" (1976), Moscow, Tsentralnaya laboratoriya okhrany prirody, 4 pp. (typescript).

Zelikman, L. (1989), "A large-scale ecological disaster is threatening from the north," *Environmental Policy Review*, 3, no. 2, pp. 1–8.

Ziegler, C. E. (1987), *Environmental Policy in the USSR*, Amherst, University of Massachusetts Press.

ZumBrunnen, C. (1984), "A review of Soviet water quality management: theory and practice," in Demko, G. and Fuchs, R. (eds.), *Geographical Studies on the Soviet Union*, Chicago, University of Chicago, ch. 13.

Zvonkova, T. V. (1984), "Natural environmental potential of the Kursk Magnetic Anomaly: trends and aspects of nature conservation," *Geoforum*, 15, no. 1, pp. 101–4.

Index

309